Coding Theory
A First Course

Coding theory is concerned with successfully transmitting data through a noisy channel and correcting errors in corrupted messages. It is of central importance for many applications in computer science or engineering. This book gives a comprehensive introduction to coding theory whilst only assuming basic linear algebra. It contains a detailed and rigorous introduction to the theory of block codes and moves on to more advanced topics such as BCH codes, Goppa codes and Sudan's algorithm for list decoding. The issues of bounds and decoding, essential to the design of good codes, feature prominently.

The authors of this book have, for several years, successfully taught a course on coding theory to students at the National University of Singapore. This book is based on their experiences and provides a thoroughly modern introduction to the subject. There is a wealth of examples and exercises, some of which introduce students to novel or more advanced material.

Coding Theory

A First Course

SAN LING
CHAOPING XING
National University of Singapore

CAMBRIDGE
UNIVERSITY PRESS

PUBLISHED BY THE PRESS SYNDICATE OF THE UNIVERSITY OF CAMBRIDGE
The Pitt Building, Trumpington Street, Cambridge, United Kingdom

CAMBRIDGE UNIVERSITY PRESS
The Edinburgh Building, Cambridge CB2 2RU, UK
40 West 20th Street, New York, NY 10011-4211, USA
477 Williamstown Road, Port Melbourne, VIC 3207, Australia
Ruiz de Alarcón 13, 28014 Madrid, Spain
Dock House, The Waterfront, Cape Town 8001, South Africa

http://www.cambridge.org

First published 2004

Printed in the United Kingdom at the University Press, Cambridge

Typeface Times 10/13 pt.   *System* LATEX 2_ε [TB]

A catalogue record for this book is available from the British Library

ISBN 0 521 82191 6 hardback
ISBN 0 521 52923 9 paperback

To Mom and Dad
and my beloved wife Bee Keow

S. L.

To my wife Youqun Shi
and my children Zhengrong and Menghong

C. P. X.

Contents

Preface

In the seminal paper 'A mathematical theory of communication' published in 1948, Claude Shannon showed that, given a noisy communication channel, there is a number, called the capacity of the channel, such that reliable communication can be achieved at any rate below the channel capacity, if proper encoding and decoding techniques are used. This marked the birth of coding theory, a field of study concerned with the transmission of data across noisy channels and the recovery of corrupted messages.

In barely more than half a century, coding theory has seen phenomenal growth. It has found widespread application in areas ranging from communication systems, to compact disc players, to storage technology. In the effort to find good codes for practical purposes, researchers have moved beyond block codes to other paradigms, such as convolutional codes, turbo codes, space-time codes, low-density-parity-check (LDPC) codes and even quantum codes. While the problems in coding theory often arise from engineering applications, it is fascinating to note the crucial role played by mathematics in the development of the field. The importance of algebra, combinatorics and geometry in coding theory is a commonly acknowledged fact, with many deep mathematical results being used in elegant ways in the advancement of coding theory.

Coding theory therefore appeals not just to engineers and computer scientists, but also to mathematicians. It has become increasingly common to find the subject taught as part of undergraduate or graduate curricula in mathematics.

This book grew out of two one-semester courses we have taught at the National University of Singapore to advanced mathematics and computer science undergraduates over a number of years. Given the vastness of the subject, we have chosen to restrict our attention to block codes, with the aim of introducing the theory without a prerequisite in algebra. The only mathematical prerequisite assumed is familiarity with basic notions and results in linear algebra. The results on finite fields needed in the book are covered in Chapter 3.

The design of good codes, from both the theoretical and practical points of view, is a very important problem in coding theory. General bounds on the parameters of codes are often used as benchmarks to determine how good a given code is, while, from the practical perspective, a code must admit an efficient decoding scheme before it can be considered useful. Since the beginning of coding theory, researchers have done much work in these directions and, in the process, have constructed many interesting families of codes. This book is built pretty much around these themes. A fairly detailed discussion on some well known bounds is included in Chapter 5, while quite a number of decoding techniques are discussed throughout this book. An effort is also made to introduce systematically many of the well known families of codes, for example, Hamming codes, Golay codes, Reed–Muller codes, cyclic codes, BCH codes, Reed–Solomon codes, alternant codes, Goppa codes, etc.

In order to stay sufficiently focused and to keep the book within a manageable size, we have to omit certain well established topics or examples, such as a thorough treatment of weight enumerators, from our discussion. Wherever possible, we try to include some of these omitted topics in the exercises at the end of each chapter. More than 250 problems have been included to help strengthen the reader's understanding and to serve as an additional source of examples and results.

Finally, it is a pleasure for us to acknowledge the help we have received while writing this book. Our research work in coding theory has received generous financial assistance from the Ministry of Education (Singapore), the National University of Singapore, the Defence Science and Technology Agency (Singapore) and the Chinese Academy of Sciences. We are thankful to these organizations for their support. We thank those who have read through the drafts carefully and provided us with invaluable feedback, especially Fangwei Fu, Wilfried Meidl, Harald Niederreiter, Yuansheng Tang (who has also offered us generous help in the preparation of Section 9.4), Arne Winterhof and Sze Ling Yeo, as well as the students in the classes MA3218 and MA4261. David Chew has been most helpful in assisting us with problems concerning LaTeX, and we are most grateful for his help. We would also like to thank Shanthi d/o Devadas for secretarial help.

1   Introduction

Information media, such as communication systems and storage devices of data, are not absolutely reliable in practice because of noise or other forms of introduced interference. One of the tasks in coding theory is to detect, or even correct, errors. Usually, coding is defined as *source coding* and *channel coding*. Source coding involves changing the message source to a suitable code to be transmitted through the channel. An example of source coding is the ASCII code, which converts each character to a byte of 8 bits. A simple communication model can be represented by Fig. 1.1.

Example 1.0.1 Consider the source encoding of four fruits, *apple*, *banana*, *cherry*, *grape*, as follows:

$$\text{apple} \rightarrow 00, \quad \text{banana} \rightarrow 01, \quad \text{cherry} \rightarrow 10, \quad \text{grape} \rightarrow 11.$$

Suppose the message 'apple', which is encoded as 00, is transmitted over a noisy channel. The message may become distorted and may be received as 01 (see Fig. 1.2). The receiver may not realize that the message was corrupted. This communication fails.

The idea of channel coding is to encode the message again after the source coding by introducing some form of redundancy so that errors can be detected or even corrected. Thus, Fig. 1.1 becomes Fig. 1.3.

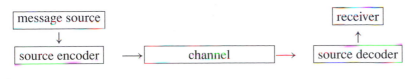

Fig. 1.1.

1

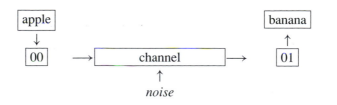

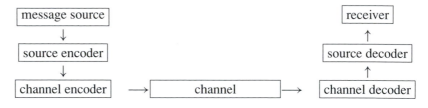

Fig. 1.2.

Fig. 1.3.

Example 1.0.2 In Example 1.0.1, we perform the channel encoding by introducing a redundancy of 1 bit as follows:

$$00 \to 000, \quad 01 \to 011, \quad 10 \to 101, \quad 11 \to 110.$$

Suppose that the message 'apple', which is encoded as 000 after the source and channel encoding, is transmitted over a noisy channel, and that there is only one error introduced. Then the received word must be one of the following three: 100, 010 or 001. In this way, we can detect the error, as none of 100, 010 or 001 is among our encoded messages.

Note that the above encoding scheme allows us to detect errors at the cost of reducing transmission speed as we have to transmit 3 bits for a message of 2 bits.

The above channel encoding scheme does not allow us to correct errors. For instance, if 100 is received, then we do not know whether 100 comes from 000, 110 or 101. However, if more redundancy is introduced, we are able to correct errors. For instance, we can design the following channel coding scheme:

$$00 \to 00000, \quad 01 \to 01111, \quad 10 \to 10110, \quad 11 \to 11001.$$

Suppose that the message 'apple' is transmitted over a noisy channel, and that there is only one error introduced. Then the received word must be one of the following five: 10000, 01000, 00100, 00010 or 00001. Assume that 10000 is received. Then we can be sure that 10000 comes from 00000 because there are

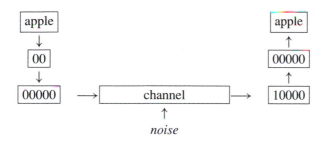

Fig. 1.4.

at least two errors between 10000 and each of the other three encoded messages 01111, 10110 and 11001.

Note that we lose even more in terms of information transmission speed in this case.

See Fig. 1.4 for this example.

Example 1.0.3 Here is a simple and general method of adding redundancy for the purpose of error correction. Assume that source coding has already been done and that the information consists of bit strings of fixed length k. Encoding is carried out by taking a bit string and repeating it $2r + 1$ times, where $r \geq 1$ is a fixed integer. For instance,

$$01 \longrightarrow 0101010101$$

if $k = 2$ and $r = 2$. In this special case, decoding is done by first considering the positions 1, 3, 5, 7, 9 of the received string and taking the first decoded bit as the one which appears more frequently at these positions; we deal similarly with the positions 2, 4, 6, 8, 10 to obtain the second decoded bit. For instance, the received string

$$1100100010$$

is decoded to 10. It is clear that, in this special case, we can decode up to two errors correctly. In the general case, we can decode up to r errors correctly. Since r is arbitrary, there are thus encoders which allow us to correct as many errors as we want. For obvious reasons, this method is called a *repetition code*. The only problem with this method is that it involves a serious loss of information transmission speed. Thus, we will look for more efficient methods.

The goal of channel coding is to construct encoders and decoders in such a way as to effect:

(1) fast encoding of messages;

(2) easy transmission of encoded messages;
(3) fast decoding of received messages;
(4) maximum transfer of information per unit time;
(5) maximal detection or correction capability.

From the mathematical point of view, the primary goals are (4) and (5). However, (5) is, in general, not compatible with (4), as we will see in Chapter 5. Therefore, any solution is necessarily a trade-off among the five objectives.

Throughout this book, we are primarily concerned with channel coding. Channel coding is also called *algebraic coding* as algebraic tools are extensively involved in the study of channel coding.

Exercises

1.1 Design a channel coding scheme to detect two or less errors for the message source {00, 10, 01, 11}. Can you find one of the best schemes in terms of information transmission speed?

1.2 Design a channel coding scheme to correct two or less errors for the message source {00, 10, 01, 11}. Can you find one of the best schemes in terms of information transmission speed?

1.3 Design a channel coding scheme to detect one error for the message source

$$\{000, 100, 010, 001, 110, 101, 011, 111\}.$$

Can you find one of the best schemes in terms of information transmission speed?

1.4 Design a channel coding scheme to correct one error for the message source

$$\{000, 100, 010, 001, 110, 101, 011, 111\}.$$

Can you find one of the best schemes in terms of information transmission speed?

2 Error detection, correction and decoding

We saw in Chapter 1 that the purpose of channel coding is to introduce redundancy to information messages so that errors that occur in the transmission can be detected or even corrected. In this chapter, we formalize and discuss the notions of error-detection and error-correction. We also introduce some well known decoding rules, i.e., methods that retrieve the original message sent by detecting and correcting the errors that have occurred in the transmission.

2.1 Communication channels

We begin with some basic definitions.

Definition 2.1.1 Let $A = \{a_1, a_2, \ldots, a_q\}$ be a set of size q, which we refer to as a *code alphabet* and whose elements are called *code symbols*.

(i) A *q-ary word* of length n over A is a sequence $\mathbf{w} = w_1 w_2 \cdots w_n$ with each $w_i \in A$ for all i. Equivalently, $\mathbf{w}$ may also be regarded as the vector $(w_1, \ldots, w_n)$.

(ii) A *q-ary block code* of length n over A is a nonempty set C of q-ary words having the same length n.

(iii) An element of C is called a *codeword* in C.

(iv) The number of codewords in C, denoted by $|C|$, is called the *size* of C.

(v) The *(information) rate* of a code C of length n is defined to be $(\log_q |C|)/n$.

(vi) A code of length n and size M is called an *(n, M)-code*.

Remark 2.1.2 In practice, and especially in this book, the code alphabet is often taken to be a finite field $\mathbf{F}_q$ of order q (cf. Chapter 3).

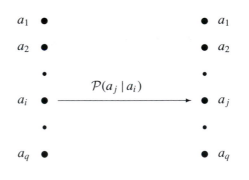

Fig. 2.1.

Example 2.1.3 A code over the code alphabet $\mathbf{F}_2 = \{0, 1\}$ is called a *binary code*; i.e., the code symbols for a binary code are 0 and 1. Some examples of binary codes are:

(i) $C_1 = \{00, 01, 10, 11\}$ is a (2,4)-code;
(ii) $C_2 = \{000, 011, 101, 110\}$ is a (3,4)-code;
(iii) $C_3 = \{0011, 0101, 1010, 1100, 1001, 0110\}$ is a (4,6)-code.

A code over the code alphabet $\mathbf{F}_3 = \{0, 1, 2\}$ is called a *ternary code*, while the term *quaternary code* is sometimes used for a code over the code alphabet $\mathbf{F}_4$. However, a code over the code alphabet $\mathbf{Z}_4 = \{0, 1, 2, 3\}$ is also sometimes referred to as a quaternary code (cf. Chapter 3 for the definitions of $\mathbf{F}_3$, $\mathbf{F}_4$ and $\mathbf{Z}_4$).

Definition 2.1.4 A *communication channel* consists of a finite *channel alphabet* $A = \{a_1, \ldots, a_q\}$ as well as a set of *forward channel probabilities* $\mathcal{P}(a_j$ received $| a_i$ sent), satisfying

$$\sum_{j=1}^{q} \mathcal{P}(a_j \text{ received} \,|\, a_i \text{ sent}) = 1$$

for all i (see Fig. 2.1). (Here, $\mathcal{P}(a_j$ received $| a_i$ sent) is the conditional probability that a_j is received, given that a_i is sent.)

Definition 2.1.5 A communication channel is said to be *memoryless* if the outcome of any one transmission is independent of the outcome of the previous transmissions; i.e., if $\mathbf{c} = c_1 c_2 \cdots c_n$ and $\mathbf{x} = x_1 x_2 \cdots x_n$ are words of length n, then

$$\mathcal{P}(\mathbf{x} \text{ received} \,|\, \mathbf{c} \text{ sent}) = \prod_{i=1}^{n} \mathcal{P}(x_i \text{ received} \,|\, c_i \text{ sent}).$$

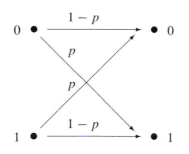

Fig. 2.2. Binary symmetric channel.

Definition 2.1.6 A *q-ary symmetric channel* is a memoryless channel which has a channel alphabet of size q such that

(i) each symbol transmitted has the same probability p $(<1/2)$ of being re-
    ceived in error;
(ii) if a symbol is received in error, then each of the $q - 1$ possible errors is
    equally likely.

In particular, the *binary symmetric channel* (*BSC*) is a memoryless channel which has channel alphabet $\{0, 1\}$ and channel probabilities

$$\mathcal{P}(1 \text{ received} \,|\, 0 \text{ sent}) = \mathcal{P}(0 \text{ received} \,|\, 1 \text{ sent}) = p,$$
$$\mathcal{P}(0 \text{ received} \,|\, 0 \text{ sent}) = \mathcal{P}(1 \text{ received} \,|\, 1 \text{ sent}) = 1 - p.$$

Thus, the probability of a bit error in a BSC is p. This is called the *crossover probability* of the BSC (see Fig. 2.2).

Example 2.1.7 Suppose that codewords from the code $\{000, 111\}$ are being sent over a BSC with crossover probability $p = 0.05$. Suppose that the word 110 is received. We can try to find the more likely codeword sent by computing the forward channel probabilities:

$$\mathcal{P}(110 \text{ received} \,|\, 000 \text{ sent}) = \mathcal{P}(1 \text{ received} \,|\, 0 \text{ sent})^2 \times \mathcal{P}(0 \text{ received} \,|\, 0 \text{ sent})$$
$$= (0.05)^2(0.95) = 0.002375,$$
$$\mathcal{P}(110 \text{ received} \,|\, 111 \text{ sent}) = \mathcal{P}(1 \text{ received} \,|\, 1 \text{ sent})^2 \times \mathcal{P}(0 \text{ received} \,|\, 1 \text{ sent})$$
$$= (0.95)^2(0.05) = 0.045125.$$

Since the second probability is larger than the first, we can conclude that 111 is more likely to be the codeword sent.

Decoding rule

In a communication channel with coding, only codewords are transmitted. Suppose that a word **w** is received. If **w** is a valid codeword, we may conclude that there is no error in the transmission. Otherwise, we know that some errors have occurred. In this case, we need a rule for finding the most likely codeword sent. Such a rule is known as a *decoding rule*. We discuss two such general rules in this chapter. Some other decoding rules, which may apply to certain specific families of codes, will also be introduced in subsequent chapters.

2.2   Maximum likelihood decoding

Suppose that codewords from a code C are being sent over a communication channel. If a word **x** is received, we can compute the forward channel probabilities

$$P(\mathbf{x} \text{ received} \mid \mathbf{c} \text{ sent})$$

for all the codewords $\mathbf{c} \in C$. The *maximum likelihood decoding (MLD) rule* will conclude that $\mathbf{c_x}$ is the most likely codeword transmitted if $\mathbf{c_x}$ maximizes the forward channel probabilities; i.e.,

$$P(\mathbf{x} \text{ received} \mid \mathbf{c_x} \text{ sent}) = \max_{\mathbf{c} \in C} P(\mathbf{x} \text{ received} \mid \mathbf{c} \text{ sent}).$$

There are two kinds of MLD:

(i) *Complete maximum likelihood decoding (CMLD)*. If a word **x** is received, find the most likely codeword transmitted. If there are more than one such codewords, select one of them arbitrarily.

(ii) *Incomplete maximum likelihood decoding (IMLD)*. If a word **x** is received, find the most likely codeword transmitted. If there are more than one such codewords, request a retransmission.

2.3   Hamming distance

Suppose that codewords from a code C are being sent over a BSC with crossover probability $p < 1/2$ (in practice, p should be much smaller than $1/2$). If a word **x** is received, then for any codeword $\mathbf{c} \in C$ the forward channel probability is given by

$$P(\mathbf{x} \text{ received} \mid \mathbf{c} \text{ sent}) = p^e (1 - p)^{n-e},$$

where n is the length of **x** and e is the number of places at which **x** and **c** differ. Since $p < 1/2$, it follows that $1 - p > p$, so this probability is larger for larger values of $n - e$, i.e., for smaller values of e. Hence, this probability is maximized by choosing a codeword $\mathbf{c}$ for which e is as small as possible. This value e leads us to introduce the following fundamental notion of Hamming distance.

Definition 2.3.1 Let $\mathbf{x}$ and $\mathbf{y}$ be words of length n over an alphabet A. The *(Hamming) distance* from $\mathbf{x}$ to $\mathbf{y}$, denoted by $d(\mathbf{x}, \mathbf{y})$, is defined to be the number of places at which $\mathbf{x}$ and $\mathbf{y}$ differ. If $\mathbf{x} = x_1 \cdots x_n$ and $\mathbf{y} = y_1 \cdots y_n$, then

$$d(\mathbf{x}, \mathbf{y}) = d(x_1, y_1) + \cdots + d(x_n, y_n), \tag{2.1}$$

where x_i and y_i are regarded as words of length 1, and

$$d(x_i, y_i) = \begin{cases} 1 & \text{if } x_i \neq y_i \\ 0 & \text{if } x_i = y_i. \end{cases}$$

Example 2.3.2 (i)   Let $A = \{0, 1\}$ and let $\mathbf{x} = 01010, \mathbf{y} = 01101, \mathbf{z} = 11101$. Then

$$d(\mathbf{x}, \mathbf{y}) = 3,$$
$$d(\mathbf{y}, \mathbf{z}) = 1,$$
$$d(\mathbf{z}, \mathbf{x}) = 4.$$

(ii)   Let $A = \{0, 1, 2, 3, 4\}$ and let $\mathbf{x} = 1234, \mathbf{y} = 1423, \mathbf{z} = 3214$. Then

$$d(\mathbf{x}, \mathbf{y}) = 3,$$
$$d(\mathbf{y}, \mathbf{z}) = 4,$$
$$d(\mathbf{z}, \mathbf{x}) = 2.$$

Proposition 2.3.3 *Let $\mathbf{x}, \mathbf{y}, \mathbf{z}$ be words of length n over A. Then we have*

(i) $0 \leq d(\mathbf{x}, \mathbf{y}) \leq n$,
(ii) $d(\mathbf{x}, \mathbf{y}) = 0$ *if and only if* $\mathbf{x} = \mathbf{y}$,
(iii) $d(\mathbf{x}, \mathbf{y}) = d(\mathbf{y}, \mathbf{x})$,
(iv) *(Triangle inequality.)* $d(\mathbf{x}, \mathbf{z}) \leq d(\mathbf{x}, \mathbf{y}) + d(\mathbf{y}, \mathbf{z})$.

Proof. (i), (ii) and (iii) are obvious from the definition of the Hamming distance. By (2.1), it is enough to prove (iv) when $n = 1$, which we now assume.
   If $\mathbf{x} = \mathbf{z}$, then (iv) is obviously true since $d(\mathbf{x}, \mathbf{z}) = 0$.
   If $\mathbf{x} \neq \mathbf{z}$, then either $\mathbf{y} \neq \mathbf{x}$ or $\mathbf{y} \neq \mathbf{z}$, so (iv) is again true. $\square$

2.4   Nearest neighbour/minimum distance decoding

Suppose that codewords from a code C are being sent over a communication channel. If a word $\mathbf{x}$ is received, the *nearest neighbour decoding rule* (or *minimum distance decoding rule*) will decode $\mathbf{x}$ to $\mathbf{c}_{\mathbf{x}}$ if $d(\mathbf{x}, \mathbf{c}_{\mathbf{x}})$ is minimal among all the codewords in C, i.e.,

$$d(\mathbf{x}, \mathbf{c}_{\mathbf{x}}) = \min_{\mathbf{c} \in C} d(\mathbf{x}, \mathbf{c}). \tag{2.2}$$

Just as for the case of maximum likelihood decoding, we can distinguish between complete and incomplete decoding for the nearest neighbour decoding rule. For a given received word $\mathbf{x}$, if two or more codewords $\mathbf{c}_{\mathbf{x}}$ satisfy (2.2), then the complete decoding rule arbitrarily selects one of them to be the most likely word sent, while the incomplete decoding rule requests for a retransmission.

Theorem 2.4.1 *For a BSC with crossover probability $p < 1/2$, the maximum likelihood decoding rule is the same as the nearest neighbour decoding rule.*

Proof. Let C denote the code in use and let $\mathbf{x}$ denote the received word (of length n). For any vector $\mathbf{c}$ of length n, and for any $0 \le i \le n$,

$$d(\mathbf{x}, \mathbf{c}) = i \Leftrightarrow \mathcal{P}(\mathbf{x} \text{ received} \,|\, \mathbf{c} \text{ sent}) = p^i (1 - p)^{n-i}.$$

Since $p < 1/2$, it follows that

$$p^0(1 - p)^n > p^1(1 - p)^{n-1} > p^2(1 - p)^{n-2} > \cdots > p^n(1 - p)^0.$$

By definition, the maximum likelihood decoding rule decodes $\mathbf{x}$ to $\mathbf{c} \in C$ such that $\mathcal{P}(\mathbf{x} \text{ received} \,|\, \mathbf{c} \text{ sent})$ is the largest, i.e., such that $d(\mathbf{x}, \mathbf{c})$ is the smallest (or seeks retransmission if incomplete decoding is in use and $\mathbf{c}$ is not unique). Hence, it is the same as the nearest neighbour decoding rule. $\qquad \square$

Remark 2.4.2 From now on, we will assume that all BSCs have crossover probabilities $p < 1/2$. Consequently, we can use the minimum distance decoding rule to perform MLD.

Example 2.4.3 Suppose codewords from the binary code

$$C = \{0000, 0011, 1000, 1100, 0001, 1001\}$$

are being sent over a BSC. Assuming $\mathbf{x} = 0111$ is received, then

$$d(0111, 0000) = 3,$$
$$d(0111, 0011) = 1,$$
$$d(0111, 1000) = 4,$$

Table 2.1. IMLD table for C.

Received $\mathbf{x}$	$d(\mathbf{x}, 000)$	$d(\mathbf{x}, 011)$	Decode to
000	0	2	000
100	1	3	000
010	1	1	–
001	1	1	–
110	2	2	–
101	2	2	–
011	2	0	011
111	3	1	011

$$d(0111, 1100) = 3,$$
$$d(0111, 0001) = 2,$$
$$d(0111, 1001) = 3.$$

By using nearest neighbour decoding, we decode $\mathbf{x}$ to 0011.

Example 2.4.4 Let $C = \{000, 011\}$ be a binary code. The IMLD table for C is as shown in Table 2.1, where '–' means that retransmission is sought.

2.5   Distance of a code

Apart from the length and size of a code, another important and useful characteristic of a code is its distance.

Definition 2.5.1 For a code C containing at least two words, the (*minimum*) *distance* of C, denoted by $d(C)$, is

$$d(C) = \min\{d(\mathbf{x}, \mathbf{y}) : \mathbf{x}, \mathbf{y} \in C, \ \mathbf{x} \neq \mathbf{y}\}.$$

Definition 2.5.2 A code of length n, size M and distance d is referred to as an (n, M, d)-*code*. The numbers n, M and d are called the *parameters* of the code.

Example 2.5.3 (i) Let $C = \{00000, 00111, 11111\}$ be a binary code. Then $d(C) = 2$ since

$$d(00000, 00111) = 3,$$
$$d(00000, 11111) = 5,$$
$$d(00111, 11111) = 2.$$

Hence, C is a binary (5,3,2)-code.

(ii) Let $C = \{000000, 000111, 111222\}$ be a ternary code (i.e. with code alphabet $\{0, 1, 2\}$). Then $d(C) = 3$ since

$$d(000000, 000111) = 3,$$
$$d(000000, 111222) = 6,$$
$$d(000111, 111222) = 6.$$

Hence, C is a ternary (6,3,3)-code.

It turns out that the distance of a code is intimately related to the error-detecting and error-correcting capabilities of the code.

Definition 2.5.4 Let u be a positive integer. A code C is *u-error-detecting* if, whenever a codeword incurs at least one but at most u errors, the resulting word is not a codeword. A code C is *exactly u-error-detecting* if it is u-error-detecting but not $(u + 1)$-error-detecting.

Example 2.5.5 (i) The binary code $C = \{00000, 00111, 11111\}$ is 1-error-detecting since changing any codeword in one position does not result in another codeword. In other words,

$$00000 \rightarrow 00111 \text{ needs to change three bits,}$$
$$00000 \rightarrow 11111 \text{ needs to change five bits,}$$
$$00111 \rightarrow 11111 \text{ needs to change two bits.}$$

In fact, C is exactly 1-error-detecting, as changing the first two positions of 00111 will result in another codeword 11111 (so C is not a 2-error-detecting code).

(ii) The ternary code $C = \{000000, 000111, 111222\}$ is 2-error-detecting since changing any codeword in one or two positions does not result in another codeword. In other words,

$$000000 \rightarrow 000111 \text{ needs to change three positions,}$$
$$000000 \rightarrow 111222 \text{ needs to change six positions,}$$
$$000111 \rightarrow 111222 \text{ needs to change six positions.}$$

In fact, C is exactly 2-error-detecting, as changing each of the last three positions of 000000 to 1 will result in the codeword 000111 (so C is not 3-error-detecting).

Theorem 2.5.6 *A code C is u-error-detecting if and only if $d(C) \geq u + 1$; i.e., a code with distance d is an exactly $(d - 1)$-error-detecting code.*

Proof. Suppose $d(C) \geq u + 1$. If $\mathbf{c} \in C$ and $\mathbf{x}$ are such that $1 \leq d(\mathbf{c}, \mathbf{x}) \leq u < d(C)$, then $\mathbf{x} \notin C$; hence, C is u-error-detecting.

On the other hand, if $d(C) < u+1$, i.e., $d(C) \leq u$, then there exist $c_1, c_2 \in C$ such that $1 \leq d(c_1, c_2) = d(C) \leq u$. It is therefore possible that we begin with c_1 and $d(C)$ errors (where $1 \leq d(C) \leq u$) are incurred such that the resulting word is c_2, another codeword in C. Hence, C is not a u-error-detecting code.   □

Remark 2.5.7 An illustration of Theorem 2.5.6 is given by comparing Examples 2.5.5 and 2.5.3.

Definition 2.5.8 Let v be a positive integer. A code C is v-*error-correcting* if minimum distance decoding is able to correct v or fewer errors, assuming that the incomplete decoding rule is used. A code C is *exactly* v-*error-correcting* if it is v-error-correcting but not $(v + 1)$-error-correcting.

Example 2.5.9 Consider the binary code $C = \{000, 111\}$. By using the minimum distance decoding rule, we see that:

- if 000 is sent and one error occurs in the transmission, then the received word (100, 010 or 001) will be decoded to 000;
- if 111 is sent and one error occurs in the transmission, then the received word (110, 101 or 011) will be decoded to 111.

In all cases, the single error has been corrected. Hence, C is 1-error-correcting.

If at least two errors occur, the decoding rule may produce the wrong codeword. For instance, if 000 is sent and 011 is received, then 011 will be decoded to 111 using the minimum distance decoding rule. Hence, C is exactly 1-error-correcting.

Theorem 2.5.10 *A code C is v-error-correcting if and only if $d(C) \geq 2v + 1$; i.e., a code with distance d is an exactly $\lfloor (d - 1)/2 \rfloor$-error-correcting code. Here, $\lfloor x \rfloor$ is the greatest integer less than or equal to x.*

Proof. '$\Leftarrow$' Suppose that $d(C) \geq 2v + 1$. Let c be the codeword sent and let x be the word received. If v or fewer errors occur in the transmission, then $d(x, c) \leq v$. Hence, for any codeword $c' \in C, c \neq c'$, we have

$$\begin{aligned} d(x, c') &\geq d(c, c') - d(x, c) \\ &\geq 2v + 1 - v \\ &= v + 1 \\ &> d(x, c). \end{aligned}$$

Thus, x will be decoded (correctly) to c if the minimum distance decoding rule is used. This shows that C is v-error-correcting.

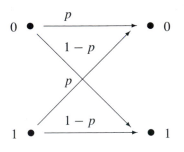

Fig. 2.3.

'⇒' Suppose that C is v-error-correcting. If $d(C) < 2v + 1$, then there are distinct codewords $\mathbf{c}, \mathbf{c}' \in C$ with $d(\mathbf{c}, \mathbf{c}') = d(C) \leq 2v$. We shall show that, assuming $\mathbf{c}$ is sent and at most v errors occur, it can occur that minimum distance decoding will either decode the received word incorrectly as $\mathbf{c}'$ or report a tie (and hence these errors cannot be corrected if the incomplete decoding rule is used). This will contradict the assumption that C is v-error-correcting, hence showing that $d(C) \geq 2v + 1$.

Notice that, if $d(\mathbf{c}, \mathbf{c}') < v + 1$, then $\mathbf{c}$ could be changed into $\mathbf{c}'$ by incurring at most v errors, and these errors would go uncorrected (in fact, undetected!) since $\mathbf{c}'$ is again in C. This, however, would contradict the assumption that C is v-error-correcting. Therefore, $d(\mathbf{c}, \mathbf{c}') \geq v + 1$. Without loss of generality, we may hence assume that $\mathbf{c}$ and $\mathbf{c}'$ differ in exactly the first $d = d(C)$ positions, where $v + 1 \leq d \leq 2v$. If the word

$$\mathbf{x} = \underbrace{x_1 \cdots\cdots x_v}_{\text{agree with } \mathbf{c}'} \underbrace{x_{v+1} \cdots\cdots x_d}_{\text{agree with } \mathbf{c}} \underbrace{x_{d+1} \cdots\cdots x_n}_{\text{agree with both}}$$

is received, then we have

$$d(\mathbf{x}, \mathbf{c}') = d - v \leq v = d(\mathbf{x}, \mathbf{c}).$$

It follows that either $d(\mathbf{x}, \mathbf{c}') < d(\mathbf{x}, \mathbf{c})$, in which case $\mathbf{x}$ is decoded incorrectly as $\mathbf{c}'$, or $d(\mathbf{x}, \mathbf{c}) = d(\mathbf{x}, \mathbf{c}')$, in which case a tie is reported. □

Exercises

2.1 Explain why the binary communication channel shown in Fig. 2.3, where $p < 0.5$, is called a *useless channel*.

2.2 Suppose that codewords from the binary code {000, 100, 111} are being sent over a BSC (binary symmetric channel) with crossover probability

$p = 0.03$. Use the maximum likelihood decoding rule to decode the following received words:

(a) 010,   (b) 011,   (c) 001.

2.3 Consider a memoryless binary channel with channel probabilities

$$P(0 \text{ received} \mid 0 \text{ sent}) = 0.7, \qquad P(1 \text{ received} \mid 1 \text{ sent}) = 0.8.$$

If codewords from the binary code $\{000, 100, 111\}$ are being sent over this channel, use the maximum likelihood decoding rule to decode the following received words:

(a) 010,   (b) 011,   (c) 001.

2.4 Let $C = \{001, 011\}$ be a binary code.

  (a) Suppose we have a memoryless binary channel with the following probabilities:

$$P(0 \text{ received} \mid 0 \text{ sent}) = 0.1 \text{ and } P(1 \text{ received} \mid 1 \text{ sent}) = 0.5.$$

  Use the maximum likelihood decoding rule to decode the received word 000.

  (b) Use the nearest neighbour decoding rule to decode 000.

2.5 For the binary code $C = \{01101, 00011, 10110, 11000\}$, use the nearest neighbour decoding rule to decode the following received words:

(a) 00000, (b) 01111, (c) 10110, (d) 10011, (e) 11011.

2.6 For the ternary code $C = \{00122, 12201, 20110, 22000\}$, use the nearest neighbour decoding rule to decode the following received words:

(a) 01122, (b) 10021, (c) 22022, (d) 20120.

2.7 Construct the IMLD (incomplete maximum likelihood decoding) table for each of the following binary codes:

  (a) $C = \{101, 111, 011\}$,
  (b) $C = \{000, 001, 010, 011\}$.

2.8 Determine the number of binary codes with parameters $(n, 2, n)$ for $n \geq 2$.

3 Finite fields

From the previous chapter, we know that a code alphabet A is a finite set. In order to play mathematical games, we are going to equip A with some algebraic structures. As we know, a field, such as the real field **R** or the complex field **C**, has two operations, namely addition and multiplication. Our idea is to define two operations for A so that A becomes a field. Of course, then A is a field with only finitely many elements, whilst **R** and **C** are fields with infinitely many elements. Fields with finitely many elements are quite different from those that we have learnt about before.

The theory of finite fields goes back to the seventeenth and eighteenth centuries, with eminent mathematicians such as Pierre de Fermat (1601–1665) and Leonhard Euler (1707–1783) contributing to the structure theory of special finite fields. The general theory of finite fields began with the work of Carl Friedrich Gauss (1777–1855) and Evariste Galois (1811–1832), but it only became of interest for applied mathematicians and engineers in recent decades because of its many applications to mathematics, computer science and communication theory. Nowadays, the theory of finite fields has become very rich. In this chapter, we only study a small portion of this theory. The reader already familiar with the elementary properties of finite fields may wish to proceed directly to the next chapter. For a more complete introduction to finite fields, the reader is invited to consult ref. [11].

3.1 Fields

Definition 3.1.1 A *field* is a nonempty set F of elements with two operations '$+$' (called addition) and '$\cdot$' (called multiplication) satisfying the following axioms. For all $a, b, c \in F$:

$$
\begin{array}{c|cc}
+ & 0 & 1 \\
\hline
0 & 0 & 1 \\
1 & 1 & 0
\end{array}
\qquad
\begin{array}{c|cc}
\times & 0 & 1 \\
\hline
0 & 0 & 0 \\
1 & 0 & 1
\end{array}
$$

Fig. 3.1. Addition and multiplication tables for $\mathbf{Z}_2$.

(i) F is closed under $+$ and $\cdot$; i.e., $a + b$ and $a \cdot b$ are in F.

(ii)   Commutative laws: $a + b = b + a$, $a \cdot b = b \cdot a$.

(iii) Associative laws: $(a + b) + c = a + (b + c)$, $a \cdot (b \cdot c) = (a \cdot b) \cdot c$.

(iv)   Distributive law: $a \cdot (b + c) = a \cdot b + a \cdot c$.

Furthermore, two distinct identity elements 0 and 1 (called the *additive* and *multiplicative identities*, respectively) must exist in F satisfying the following:

(v) $a + 0 = a$ for all $a \in F$.

(vi) $a \cdot 1 = a$ and $a \cdot 0 = 0$ for all $a \in F$.

(vii) For any a in F, there exists an additive inverse element $(-a)$ in F such that $a + (-a) = 0$.

(viii) For any $a \neq 0$ in F, there exists a multiplicative inverse element a^{-1} in F such that $a \cdot a^{-1} = 1$.

We usually write $a \cdot b$ simply as ab, and denote by F^* the set $F \backslash \{0\}$.

Example 3.1.2 (i) Some familiar fields are the rational field

$$
\mathbf{Q} := \left\{ \frac{a}{b} : a, b \text{ are integers with } b \neq 0 \right\},
$$

the real field $\mathbf{R}$ and the complex field $\mathbf{C}$. It is easy to check that all the axioms in Definition 3.1.1 are satisfied for the above three fields. In fact, we are not interested in these fields because all of them have an infinite number of elements.

(ii) Denote by $\mathbf{Z}_2$ the set $\{0, 1\}$. We define the addition and multiplication as in Fig. 3.1.

Then, it is easy to check that $\mathbf{Z}_2$ is a field. It has only two elements!

More properties of a field can be deduced from the definition.

Lemma 3.1.3 *Let a, b be any two elements of a field F. Then*

(i) $(-1) \cdot a = -a$;

(ii) $ab = 0$ *implies* $a = 0$ *or* $b = 0$.

Proof. (i) We have

$$(-1) \cdot a + a \stackrel{\text{(vi)}}{=} (-1) \cdot a + a \cdot 1 \stackrel{\text{(ii),(iv)}}{=} ((-1)+1) \cdot a \stackrel{\text{(vii)}}{=} 0 \cdot a \stackrel{\text{(ii),(vi)}}{=} 0,$$

where the Roman numerals in the above formula stand for the axioms in Definition 3.1.1. Thus, $(-1) \cdot a = -a$.

(ii) If $a \neq 0$, then

$$0 \stackrel{\text{(vi)}}{=} a^{-1} \cdot 0 = a^{-1}(ab) \stackrel{\text{(iii)}}{=} (a^{-1}a)b \stackrel{\text{(ii),(viii)}}{=} 1 \cdot b \stackrel{\text{(ii)}}{=} b \cdot 1 \stackrel{\text{(vi)}}{=} b,$$

where the Roman numerals in the above formula again stand for the axioms in Definition 3.1.1.   □

A field containing only finitely many elements is called a *finite field*. A set F satisfying axioms (i)–(vii) in Definition 3.1.1 is called a *(commutative) ring*.

Example 3.1.4 (i) The set of all integers

$$\mathbf{Z} := \{0, \pm 1, \pm 2, \ldots\}$$

forms a ring under the normal addition and multiplication. It is called the *integer ring*.

(ii) The set of all polynomials over a field F,

$$F[x] := \{a_0 + a_1 x + \cdots + a_n x^n : a_i \in F, n \geq 0\},$$

forms a ring under the normal addition and multiplication of polynomials.

Definition 3.1.5 Let a, b and $m > 1$ be integers. We say that a is *congruent to b modulo m*, written as

$$a \equiv b \pmod{m},$$

if $m|(a - b)$; i.e., m divides $a - b$.

Example 3.1.6

(i) $90 \equiv 30 \pmod{60}$ and $15 \equiv 3 \pmod{12}$.
(ii) $a \equiv 0 \pmod{m}$ means that $m|a$.
(iii) $a \equiv 0 \pmod{2}$ means that a is even.
(iv) $a \equiv 1 \pmod{2}$ means that a is odd.

Remark 3.1.7 Given integers a and $m > 1$, by the division algorithm we have

$$a = mq + b, \tag{3.1}$$

+	0 1 2 3
0	0 1 2 3
1	1 2 3 0
2	2 3 0 1
3	3 0 1 2

·	0 1 2 3
0	0 0 0 0
1	0 1 2 3
2	0 2 0 2
3	0 3 2 1

Fig. 3.2. Addition and multiplication tables for $\mathbf{Z}_4$.

where b is uniquely determined by a and m, and $0 \leq b \leq m - 1$. Hence, any integer a is congruent to exactly one of $0, 1, \ldots, m - 1$ modulo m. The integer b in (3.1) is called the (*principal*) remainder of a divided by m, denoted by $(a \,(\mathrm{mod}\, m))$.

If $a \equiv b \,(\mathrm{mod}\, m)$ and $c \equiv d \,(\mathrm{mod}\, m)$, then we have

$$a + c \equiv b + d \,(\mathrm{mod}\, m),$$

$$a - c \equiv b - d \,(\mathrm{mod}\, m),$$

$$a \times c \equiv b \times d \,(\mathrm{mod}\, m).$$

For an integer $m > 1$, we denote by $\mathbf{Z}_m$ or $\mathbf{Z}/(m)$ the set $\{0, 1, \ldots, m - 1\}$ and define the addition $\oplus$ and multiplication $\odot$ in $\mathbf{Z}_m$ by:

$a \oplus b =$ the remainder of $a + b$ divided by m, i.e., $(a + b \,(\mathrm{mod}\, m))$,

and

$a \odot b =$ the remainder of ab divided by m, i.e., $(ab \,(\mathrm{mod}\, m))$.

It is easy to show that all the axioms (i)–(vii) in Definition 3.1.1 are satisfied. Hence, $\mathbf{Z}_m$, together with the addition $\oplus$ and multiplication $\odot$ defined above, forms a ring.

We will continue to denote '$\oplus$' and '$\odot$' in $\mathbf{Z}_m$ by '$+$' and '$\cdot$', respectively.

Example 3.1.8 (i) Modulo 2: the field $\mathbf{Z}_2$ in Example 3.1.2(ii) is exactly the ring defined above for $m = 2$. In this case, axiom (viii) is also satisfied. Thus, it is a field.

(ii) Modulo 4: we construct the addition and multiplication tables for $\mathbf{Z}_4$ (Fig. 3.2). From the multiplication table in Fig. 3.2, we can see that $\mathbf{Z}_4$ is not a field since 2^{-1} does not exist.

We find from the above example that $\mathbf{Z}_m$ is a field for some integers m and is just a ring for other integers. In fact, we have the following pleasing result.

Theorem 3.1.9 $\mathbf{Z}_m$ *is a field if and only if m is a prime.*

Proof. Suppose that m is a composite number and let $m = ab$ for two integers $1 < a, b < m$. Thus, $a \neq 0, b \neq 0$. However, $0 = m = a \cdot b$ in $\mathbf{Z}_m$. This is a contradiction to Lemma 3.1.3(ii). Hence, $\mathbf{Z}_m$ is not a field.

Now let m be a prime. For any nonzero element $a \in \mathbf{Z}_m$, i.e., $0 < a < m$, we know that a is prime to m. Thus, there exist two integers u, v with $0 \leq u \leq m-1$ such that $ua + vm = 1$, i.e., $ua \equiv 1 \pmod{m}$. Hence, $u = a^{-1}$. This implies that axiom (viii) in Definition 3.1.1 is also satisfied and hence $\mathbf{Z}_m$ is a field. $\square$

For a ring R, an integer $n \geq 1$ and $a \in R$, we denote by na or $n \cdot a$ the element

$$\sum_{i=1}^{n} a = \underbrace{a + a + \cdots + a}_{n}.$$

Definition 3.1.10 Let F be a field. The *characteristic* of F is the least positive integer p such that $p \cdot 1 = 0$, where 1 is the multiplicative identity of F. If no such p exists, we define the characteristic to be 0.

Example 3.1.11 (i) The characteristics of $\mathbf{Q, R, C}$ are 0.

(ii) The characteristic of the field $\mathbf{Z}_p$ is p for any prime p.

It follows from the following result that the characteristic of a field cannot be composite.

Theorem 3.1.12 *The characteristic of a field is either* 0 *or a prime number.*

Proof. It is clear that 1 is not the characteristic as $1 \cdot 1 = 1 \neq 0$.

Suppose that the characteristic p of a field F is composite. Let $p = nm$ for some positive integers $1 < n, m < p$. Put $a = n \cdot 1$ and $b = m \cdot 1$, where 1 is the multiplicative identity of F. Then,

$$a \cdot b = (n \cdot 1)(m \cdot 1) = \left(\sum_{i=1}^{n} 1\right)\left(\sum_{j=1}^{m} 1\right) = (mn) \cdot 1 = p \cdot 1 = 0.$$

By Lemma 3.1.3(ii), $a = 0$ or $b = 0$; i.e., $m \cdot 1 = 0$ or $n \cdot 1 = 0$. This contradicts the definition of the characteristic. $\square$

Let E, F be two fields and let F be a subset of E. The field F is called a *subfield* of E if the addition and multiplication of E, when restricted to F, are the same as those of F.

Example 3.1.13 (i) The rational number field $\mathbf{Q}$ is a subfield of both the real field $\mathbf{R}$ and the complex field $\mathbf{C}$, and $\mathbf{R}$ is a subfield of $\mathbf{C}$.

(ii) Let F be a field of characteristic p; then, $\mathbf{Z}_p$ can be naturally viewed as a subfield of F.

Theorem 3.1.14 *A finite field F of characteristic p contains p^n elements for some integer $n \geq 1$.*

Proof. Choose an element α_1 from F^*. We claim that $0 \cdot \alpha_1, 1 \cdot \alpha_1, \ldots, (p-1) \cdot \alpha_1$ are pairwise distinct. Indeed, if $i \cdot \alpha_1 = j \cdot \alpha_1$ for some $0 \leq i \leq j \leq p-1$, then $(j-i) \cdot \alpha_1 = 0$ and $0 \leq j-i \leq p-1$. As the characteristic of F is p, this forces $j-i = 0$; i.e., $i = j$.

If $F = \{0 \cdot \alpha_1, 1 \cdot \alpha_1, \ldots, (p-1) \cdot \alpha_1\}$, we are done. Otherwise, we choose an element α_2 in $F \backslash \{0 \cdot \alpha_1, 1 \cdot \alpha_1, \ldots, (p-1) \cdot \alpha_1\}$. We claim that $a_1\alpha_1 + a_2\alpha_2$ are pairwise distinct for all $0 \leq a_1, a_2 \leq p-1$. Indeed, if

$$a_1\alpha_1 + a_2\alpha_2 = b_1\alpha_1 + b_2\alpha_2 \tag{3.2}$$

for some $0 \leq a_1, a_2, b_1, b_2 \leq p-1$, then we must have $a_2 = b_2$. Otherwise, we would have from (3.2) that $\alpha_2 = (b_2 - a_2)^{-1}(a_1 - b_1)\alpha_1$. This is a contradiction to our choice of α_2. Since $a_2 = b_2$, it follows immediately from (3.2) that $(a_1, a_2) = (b_1, b_2)$. As F has only finitely many elements, we can continue in this fashion and obtain elements $\alpha_1, \ldots, \alpha_n$ such that

$$\alpha_i \in F \backslash \{a_1\alpha_1 + \cdots + a_{i-1}\alpha_{i-1} : a_1, \ldots, a_{i-1} \in \mathbf{Z}_p\} \quad \text{for all } 2 \leq i \leq n,$$

and

$$F = \{a_1\alpha_1 + \cdots + a_n\alpha_n : a_1, \ldots, a_n \in \mathbf{Z}_p\}.$$

In the same manner, we can show that $a_1\alpha_1 + \cdots + a_n\alpha_n$ are pairwise distinct for all $a_i \in \mathbf{Z}_p, i = 1, \ldots, n$. This implies that $|F| = p^n$. ☐

3.2 Polynomial rings

Definition 3.2.1 Let F be a field. The set

$$F[x] := \left\{ \sum_{i=0}^{n} a_i x^i : a_i \in F, n \geq 0 \right\}$$

is called the *polynomial ring* over F. (F is indeed a ring, namely axioms (i)–(vii) of Definition 3.1.1 are satisfied.) An element of $F[x]$ is called a *polynomial* over F. For a polynomial $f(x) = \sum_{i=0}^{n} a_i x^i$, the integer n is called the *degree* of $f(x)$, denoted by $\deg(f(x))$, if $a_n \neq 0$ (for convenience, we define $\deg(0) = -\infty$). Furthermore, a nonzero polynomial $f(x) = \sum_{i=0}^n a_i x^i$ of degree n is said to be *monic* if $a_n = 1$. A polynomial $f(x)$ of positive degree is said to be *reducible* (over F) if there exist two polynomials $g(x)$ and $h(x)$ over F such that $\deg(g(x)) < \deg(f(x))$, $\deg(h(x)) < \deg(f(x))$ and $f(x) = g(x)h(x)$. Otherwise, the polynomial $f(x)$ of positive degree is said to be *irreducible* (over F).

Example 3.2.2 (i) The polynomial $f(x) = x^4 + 2x^6 \in \mathbf{Z}_3[x]$ is of degree 6. It is reducible as $f(x) = x^4(1 + 2x^2)$.

(ii) The polynomial $g(x) = 1 + x + x^2 \in \mathbf{Z}_2[x]$ is of degree 2. It is irreducible. Otherwise, it would have a linear factor x or $x + 1$; i.e., 0 or 1 would be a root of $g(x)$, but $g(0) = g(1) = 1 \in \mathbf{Z}_2$.

(iii) Using the same arguments as in (ii), we can show that both $1 + x + x^3$ and $1 + x^2 + x^3$ are irreducible over $\mathbf{Z}_2$ as they have no linear factors.

Definition 3.2.3 Let $f(x) \in F[x]$ be a polynomial of degree $n \geq 1$. Then, for any polynomial $g(x) \in F[x]$, there exists a unique pair $(s(x), r(x))$ of polynomials with $\deg(r(x)) < \deg(f(x))$ or $r(x) = 0$ such that $g(x) = s(x)f(x) + r(x)$. The polynomial $r(x)$ is called the (*principal*) *remainder* of $g(x)$ divided by $f(x)$, denoted by $(g(x) \pmod{f(x)})$.

For example, let $f(x) = 1 + x^2$ and $g(x) = x + 2x^4$ be two polynomials in $\mathbf{Z}_5[x]$. Since we have $g(x) = x + 2x^4 = (3 + 2x^2)(1 + x^2) + (2 + x) = (3 + 2x^2)f(x) + (2 + x)$, the remainder of $g(x)$ divided by $f(x)$ is $2 + x$.

Analogous to the integral ring $\mathbf{Z}$, we can introduce the following notions.

Definition 3.2.4 Let $f(x), g(x) \in F[x]$ be two nonzero polynomials. The *greatest common divisor* of $f(x), g(x)$, denoted by $\gcd(f(x), g(x))$, is the monic polynomial of the highest degree which is a divisor of both $f(x)$ and $g(x)$. In particular, we say that $f(x)$ is *co-prime* (or *prime*) to $g(x)$ if $\gcd(f(x), g(x)) = 1$. The *least common multiple* of $f(x), g(x)$, denoted by $\mathrm{lcm}(f(x), g(x))$, is the monic polynomial of the lowest degree which is a multiple of both $f(x)$ and $g(x)$.

Remark 3.2.5 (i) If $f(x)$ and $g(x)$ have the following factorizations:

$$f(x) = a \cdot p_1(x)^{e_1} \cdots p_n(x)^{e_n}, \quad g(x) = b \cdot p_1(x)^{d_1} \cdots p_n(x)^{d_n},$$

where $a, b \in F^*$, $e_i, d_i \geq 0$ and $p_i(x)$ are distinct monic irreducible polynomials (the existence and uniqueness of such a polynomial factorization are

Table 3.1. Analogies between $\mathbf{Z}$ and $F[x]$.

The integral ring $\mathbf{Z}$	The polynomial ring $F[x]$
An integer m	A polynomial $f(x)$
A prime number p	An irreducible polynomial $p(x)$

Table 3.2. More analogies between $\mathbf{Z}$ and $F[x]$.

$\mathbf{Z}_m = \{0, 1, \ldots, m-1\}$	$F[x]/(f(x)) := \{\sum_{i=0}^{n-1} a_i x^i : a_i \in F, n \geq 1\}$
$a \oplus b := (a + b \,(\mathrm{mod}\, m))$	$g(x) \oplus h(x) := (g(x) + h(x) \,(\mathrm{mod}\, f(x)))$
$a \odot b := (ab \,(\mathrm{mod}\, m))$	$g(x) \odot h(x) := (g(x)h(x) \,(\mathrm{mod}\, f(x)))$
$\mathbf{Z}_m$ is a ring	$F[x]/(f(x))$ is a ring
$\mathbf{Z}_m$ is a field $\Leftrightarrow m$ is a prime	$F[x]/(f(x))$ is a field $\Leftrightarrow f(x)$ is irreducible

well-known facts, cf. Theorem 1.59 of ref. [11]), then

$$\gcd(f(x), g(x)) = p_1(x)^{\min\{e_1, d_1\}} \cdots p_n(x)^{\min\{e_n, d_n\}}$$

and

$$\mathrm{lcm}(f(x), g(x)) = p_1(x)^{\max\{e_1, d_1\}} \cdots p_n(x)^{\max\{e_n, d_n\}}.$$

(ii) Let $f(x), g(x) \in F[x]$ be two nonzero polynomials. Then there exist two polynomials $u(x), v(x)$ with $\deg(u(x)) < \deg(g(x))$ and $\deg(v(x)) < \deg(f(x))$ such that

$$\gcd(f(x), g(x)) = u(x)f(x) + v(x)g(x).$$

(iii) From (ii), it is easily shown that $\gcd(f(x)h(x), g(x)) = \gcd(f(x), g(x))$ if $\gcd(h(x), g(x)) = 1$.

There are many analogies between the integral ring $\mathbf{Z}$ and a polynomial ring $F[x]$. We list some of them in Table 3.1.

Apart from the analogies in Table 3.1, we have the division algorithm, greatest common divisors, least common multiples, etc., in both rings. Since, for each integer $m > 1$ of $\mathbf{Z}$, the ring $\mathbf{Z}_m = \mathbf{Z}/(m)$ is constructed, we can guess that the ring, denoted by $F[x]/(f(x))$, can be constructed for a given polynomial $f(x)$ of degree $n \geq 1$. We make up Table 3.2 to define the ring $F[x]/(f(x))$ and compare it with $\mathbf{Z}/(m)$.

We list the last two statements in the second column of Table 3.2 as a theorem.

+	0	1	x	$1+x$
0	0	1	x	$1+x$
1	1	0	$1+x$	x
x	x	$1+x$	0	1
$1+x$	$1+x$	x	1	0

$\times$	0	1	x	$1+x$
0	0	0	0	0
1	0	1	x	$1+x$
x	0	x	1	$1+x$
$1+x$	0	$1+x$	$1+x$	0

Fig. 3.3. Addition and multiplication tables for $\mathbf{Z}_2[x]/(1+x^2)$.

Theorem 3.2.6 *Let $f(x)$ be a polynomial over a field F of degree ≥ 1. Then $F[x]/(f(x))$, together with the addition and multiplication defined in Table 3.2, forms a ring. Furthermore, $F[x]/(f(x))$ is a field if and only if $f(x)$ is irreducible.*

Proof. It is easy to verify that $F[x]/(f(x))$ is a ring. By applying exactly the same arguments as in the proof of Theorem 3.1.9, we can prove the second part. $\square$

Remark 3.2.7 (i) We will still denote '$\oplus$' and '$\odot$' in $F[x]/(f(x))$ by '$+$' and '$\cdot$', respectively.

(ii) If $f(x)$ is a linear polynomial, then the field $F[x]/(f(x))$ is the field F itself.

Example 3.2.8 (i) Consider the ring $\mathbf{R}[x]/(1+x^2) = \{a + bx : a, b \in \mathbf{R}\}$. It is a field since $1 + x^2$ is irreducible over $\mathbf{R}$. In fact, it is the complex field $\mathbf{C}$! To see this, we just replace x in $\mathbf{R}[x]/(1+x^2)$ by the imaginary unit i.

(ii) Consider the ring

$$\mathbf{Z}_2[x]/(1+x^2) = \{0, 1, x, 1+x\}.$$

We construct the addition and multiplication tables as shown in Fig. 3.3. We see from the multiplication table in Fig. 3.3 that $\mathbf{Z}_2[x]/(1+x^2)$ is not a field as $(1+x)(1+x) = 0$.

(iii) Consider the ring

$$\mathbf{Z}_2[x]/(1+x+x^2) = \{0, 1, x, 1+x\}.$$

As $1 + x + x^2$ is irreducible over $\mathbf{Z}_2$, the ring $\mathbf{Z}_2[x]/(1+x+x^2)$ is in fact a field. This can also be verified by the addition and multiplication tables in Fig. 3.4.

+	0	1	x	$1+x$
0	0	1	x	$1+x$
1	1	0	$1+x$	x
x	x	$1+x$	0	1
$1+x$	$1+x$	x	1	0

×	0	1	x	$1+x$
0	0	0	0	0
1	0	1	x	$1+x$
x	0	x	$1+x$	1
$1+x$	0	$1+x$	1	x

Fig. 3.4. Addition and multiplication tables for $\mathbf{Z}_2[x]/(1+x+x^2)$.

3.3 Structure of finite fields

Lemma 3.3.1 *For every element β of a finite field F with q elements, we have $\beta^q = \beta$.*

Proof. It is trivial for the case where β is zero. Now assume that $\beta \neq 0$. We label all the nonzero elements of F: $F^* = \{\beta_1, \ldots, \beta_{q-1}\}$. Thus, $F^* = \{\beta\beta_1, \ldots, \beta\beta_{q-1}\}$. We obtain $\beta_1 \cdots \beta_{q-1} = (\beta\beta_1)\cdots(\beta\beta_{q-1})$, i.e., $\beta_1 \cdots \beta_{q-1} = (\beta)^{q-1}(\beta_1 \cdots \beta_{q-1})$. Hence, $\beta^{q-1} = 1$. The desired result follows. $\square$

Corollary 3.3.2 *Let F be a subfield of E with $|F| = q$. Then an element β of E lies in F if and only if $\beta^q = \beta$.*

Proof. '$\Rightarrow$' This is clear from Lemma 3.3.1.

'$\Leftarrow$' Consider the polynomial $x^q - x$. It has at most q distinct roots in E (see Theorem 1.66 of ref. [11]). As all the elements of F are roots of $x^q - x$, and $|F| = q$, we obtain $F = \{$all roots of $x^q - x$ in $E\}$. Hence, for any $\beta \in E$ satisfying $\beta^q = \beta$, it is a root of $x^q - x$; i.e., β lies in F. $\square$

For a field F of characteristic $p > 0$, we can easily show that $(\alpha + \beta)^{p^m} = \alpha^{p^m} + \beta^{p^m}$ for any $\alpha, \beta \in F$ and $m \geq 0$ (see Exercise 3.4(iii)).

For two fields E and F, the composite field $E \cdot F$ is the smallest field containing both E and F.

Using these results, we are ready to prove the main characterization of finite fields.

Theorem 3.3.3 *For any prime p and integer $n \geq 1$, there exists a unique finite field of p^n elements.*

Proof. (Existence) Let $f(x)$ be an irreducible polynomial over $\mathbf{Z}_p$ (note that the existence of such a polynomial is guaranteed by Exercise 3.28(ii) by showing that $I_p(n) > 0$ for all primes p and integers $n > 0$). It follows from Theorem 3.2.6 that the residue ring $\mathbf{Z}_p[x]/(f(x))$ is in fact a field. It is easy to verify that this field has exactly p^n elements.

(Uniqueness) Let E and F be two fields of p^n elements. In the composite field $E \cdot F$, consider the polynomial $x^{p^n} - x$ over $E \cdot F$. By Corollary 3.3.2, $E = \{\text{all roots of } x^{p^n} - x\} = F$.   □

From now on, it makes sense to denote the finite field with q elements by $\mathbf{F}_q$ or $GF(q)$.

For an irreducible polynomial $f(x)$ of degree n over a field F, let α be a root of $f(x)$. Then the field $F[x]/(f(x))$ can be represented as

$$F[\alpha] = \{a_0 + a_1\alpha + \cdots + a_{n-1}\alpha^{n-1} : a_i \in F\} \qquad (3.3)$$

if we replace x in $F[x]/(f(x))$ by α, as we did in Example 3.2.8(i). An advantage of using $F[\alpha]$ to replace the field $F[x]/(f(x))$ is that we can avoid the confusion between an element of $F[x]/(f(x))$ and a polynomial over F.

Definition 3.3.4 An element α in a finite field $\mathbf{F}_q$ is called a *primitive element* (or *generator*) of $\mathbf{F}_q$ if $\mathbf{F}_q = \{0, \alpha, \alpha^2, \ldots, \alpha^{q-1}\}$.

Example 3.3.5 Consider the field $\mathbf{F}_4 = \mathbf{F}_2[\alpha]$, where α is a root of the irreducible polynomial $1 + x + x^2 \in \mathbf{F}_2[x]$. Then we have

$$\alpha^2 = -(1+\alpha) = 1+\alpha, \quad \alpha^3 = \alpha(\alpha^2) = \alpha(1+\alpha) = \alpha+\alpha^2 = \alpha+1+\alpha = 1.$$

Thus, $\mathbf{F}_4 = \{0, \alpha, 1 + \alpha, 1\} = \{0, \alpha, \alpha^2, \alpha^3\}$, so α is a primitive element.

Definition 3.3.6 The *order* of a nonzero element $\alpha \in \mathbf{F}_q$, denoted by ord(α), is the smallest positive integer k such that $\alpha^k = 1$.

Example 3.3.7 Since there are no linear factors for the polynomial $1 + x^2$ over $\mathbf{F}_3$, $1 + x^2$ is irreducible over $\mathbf{F}_3$. Consider the element α in the field $\mathbf{F}_9 = \mathbf{F}_3[\alpha]$, where α is a root of $1 + x^2$. Then $\alpha^2 = -1$, $\alpha^3 = \alpha(\alpha^2) = -\alpha$ and

$$\alpha^4 = (\alpha^2)^2 = (-1)^2 = 1.$$

This means that ord(α) = 4.

Lemma 3.3.8 (i) *The order* ord(α) *divides* $q - 1$ *for every* $\alpha \in \mathbf{F}_q^*$.
  (ii) *For two nonzero elements* $\alpha, \beta \in \mathbf{F}_q^*$, *if* gcd(ord($\alpha$), ord($\beta$)) $= 1$, *then* ord($\alpha\beta$) = ord(α) × ord(β).

Proof. (i) Let m be a positive integer satisfying $\alpha^m = 1$. Write $m = a \cdot \text{ord}(\alpha) + b$ for some integers $a \geq 0$ and $0 \leq b < \text{ord}(\alpha)$. Then

$$1 = \alpha^m = \alpha^{a \cdot \text{ord}(\alpha)+b} = (\alpha^{\text{ord}(\alpha)})^a \cdot \alpha^b = \alpha^b.$$

This forces $b = 0$; i.e., $\text{ord}(\alpha)$ is a divisor of m. Since $\alpha^{q-1} = 1$, we obtain $\text{ord}(\alpha)|(q-1)$.

(ii) Put $r = \text{ord}(\alpha) \times \text{ord}(\beta)$. It is clear that $\alpha^r = 1 = \beta^r$ as both $\text{ord}(\alpha)$ and $\text{ord}(\beta)$ are divisors of r. Thus, $(\alpha\beta)^r = \alpha^r\beta^r = 1$. Therefore, $\text{ord}(\alpha\beta) \le \text{ord}(\alpha) \times \text{ord}(\beta)$. On the other hand, put $t = \text{ord}(\alpha\beta)$. We have

$$1 = (\alpha\beta)^{t \cdot \text{ord}(\alpha)} = (\alpha^{\text{ord}(\alpha)})^t \beta^{t \cdot \text{ord}(\alpha)} = \beta^{t \cdot \text{ord}(\alpha)}.$$

This implies that $\text{ord}(\beta)$ divides $t \cdot \text{ord}(\alpha)$ by the proof of part (i), so $\text{ord}(\beta)$ divides t as $\text{ord}(\alpha)$ is prime to $\text{ord}(\beta)$. In the same way, we can show that $\text{ord}(\alpha)$ divides t. This implies that $\text{ord}(\alpha) \times \text{ord}(\beta)$ divides t. Thus, $\text{ord}(\alpha\beta) = t \ge \text{ord}(\alpha) \times \text{ord}(\beta)$. The desired result follows. □

We now show the existence of primitive elements and give a characterization of primitive elements in terms of their order.

Proposition 3.3.9 (i) *A nonzero element of* $\mathbf{F}_q$ *is a primitive element if and only if its order is* $q - 1$.

(ii) *Every finite field has at least one primitive element.*

Proof. (i) It is easy to see that $\alpha \in \mathbf{F}_q^*$ has order $q - 1$ if and only if the elements $\alpha, \alpha^2, \ldots, \alpha^{q-1}$ are distinct. This is equivalent to saying that $\mathbf{F}_q = \{0, \alpha, \alpha^2, \ldots, \alpha^{q-1}\}$.

(ii) Let m be the least common multiple of the orders of all the elements of $\mathbf{F}_q^*$. If r^k is a prime power in the canonical factorization of m, then $r^k|\text{ord}(\alpha)$ for some $\alpha \in \mathbf{F}_q^*$. The order of $\alpha^{\text{ord}(\alpha)/r^k}$ is r^k. Thus, if

$$m = r_1^{k_1} \cdots r_n^{k_n}$$

is the canonical factorization of m for distinct primes $r_1, \ldots, r_n$, then for each $i = 1, \ldots, n$ there exists $\beta_i \in \mathbf{F}_q^*$ with $\text{ord}(\beta_i) = r_i^{k_i}$. Lemma 3.3.8(ii) implies that there exists $\beta \in \mathbf{F}_q^*$ with $\text{ord}(\beta) = m$. Now $m|(q - 1)$ by Lemma 3.3.8(i), and, on the other hand, all the $q - 1$ elements of $\mathbf{F}_q^*$ are roots of the polynomial $x^m - 1$, so that $m \ge q - 1$. Hence, $\text{ord}(\beta) = m = q - 1$, and the result follows from part (i). □

Remark 3.3.10 (i) Primitive elements are not unique. This can be seen from Exercises 3.12–3.15.

(ii) If α is a root of an irreducible polynomial of degree m over $\mathbf{F}_q$, and it is also a primitive element of $\mathbf{F}_{q^m} = \mathbf{F}_q[\alpha]$, then every element in $\mathbf{F}_{q^m}$ can be represented both as a polynomial in α and as a power of α, since $\mathbf{F}_{q^m} = \{a_0 + a_1\alpha + \cdots + a_{m-1}\alpha^{m-1} : a_i \in \mathbf{F}_q\} = \{0, \alpha, \alpha^2, \ldots, \alpha^{q^m-1}\}$. Addition for the elements of $\mathbf{F}_{q^m}$ is easily carried out if the elements are represented

Table 3.3. Elements of $\mathbf{F}_8$.

$0 = 0$	$1 = \alpha^7 = \alpha^0$	$\alpha = \alpha^1$	$\alpha^2 = \alpha^2$
$1 + \alpha = \alpha^3$	$\alpha + \alpha^2 = \alpha^4$	$1 + \alpha + \alpha^2 = \alpha^5$	$1 + \alpha^2 = \alpha^6$

as polynomials in α, whilst multiplication is easily done if the elements are represented as powers of α.

Example 3.3.11 Let α be a root of $1 + x + x^3 \in \mathbf{F}_2[x]$ (by Example 3.2.2(iii), this polynomial is irreducible over $\mathbf{F}_2$). Hence, $\mathbf{F}_8 = \mathbf{F}_2[\alpha]$. The order of α is a divisor of $8 - 1 = 7$. Thus, $\mathrm{ord}(\alpha) = 7$ and α is a primitive element. In fact, any nonzero element in $\mathbf{F}_8$ except 1 is a primitive element. We list a table (see Table 3.3) for the elements of $\mathbf{F}_8$ expressed in two forms.

Based on Table 3.3, both the addition and multiplication in $\mathbf{F}_8$ can be easily implemented. We use powers of α to represent the elements in $\mathbf{F}_8$. For instance,

$$\alpha^3 + \alpha^6 = (1 + \alpha) + (1 + \alpha^2) = \alpha + \alpha^2 = \alpha^4, \quad \alpha^3 \cdot \alpha^6 = \alpha^9 = \alpha^2.$$

From the above example, we know that both the addition and multiplication can be carried out easily if we have a table representing the elements of finite fields both in polynomial form and as powers of primitive elements. In fact, we can simplify this table by using another table, called *Zech's log table*, constructed as follows.

Let α be a primitive element of $\mathbf{F}_q$. For each $0 \le i \le q - 2$ or $i = \infty$, we determine and tabulate $z(i)$ such that $1 + \alpha^i = \alpha^{z(i)}$ (note that we set $\alpha^\infty = 0$). Then for any two elements α^i and α^j with $0 \le i \le j \le q - 2$ in $\mathbf{F}_q$, we obtain

$$\alpha^i + \alpha^j = \alpha^i(1 + \alpha^{j-i}) = \alpha^{i+z(j-i) \ (\mathrm{mod} \ q-1)}, \quad \alpha^i \cdot \alpha^j = \alpha^{i+j \ (\mathrm{mod} \ q-1)}.$$

Example 3.3.12 Let α be a root of $1 + 2x + x^3 \in \mathbf{F}_3[x]$. This polynomial is irreducible over $\mathbf{F}_3$ as it has no linear factors. Hence, $\mathbf{F}_{27} = \mathbf{F}_3[\alpha]$. The order of α is a divisor of $27 - 1 = 26$. Thus, $\mathrm{ord}(\alpha)$ is 2, 13 or 26. First, $\mathrm{ord}(\alpha) \ne 2$; otherwise, α would be 1 or -1, neither of which is a root of $1 + 2x + x^3$. Furthermore, we have $\alpha^{13} = -1 \ne 1$. Thus, $\mathrm{ord}(\alpha) = 26$ and α is a primitive element of $\mathbf{F}_{27}$. After some computation, we obtain a Zech's log table for $\mathbf{F}_{27}$ with respect to α (Table 3.4). Now we can carry out operations in $\mathbf{F}_{27}$ easily. For instance, we have

$$\alpha^7 + \alpha^{11} = \alpha^7(1 + \alpha^4) = \alpha^7 \cdot \alpha^{18} = \alpha^{25}, \quad \alpha^7 \cdot \alpha^{11} = \alpha^{18}.$$

Table 3.4. Zech's log table for $\mathbf{F}_{27}$.

i	$z(i)$	i	$z(i)$	i	$z(i)$
∞	0	8	15	17	20
0	13	9	3	18	7
1	9	10	6	19	23
2	21	11	10	20	5
3	1	12	2	21	12
4	18	13	∞	22	14
5	17	14	16	23	24
6	11	15	25	24	19
7	4	16	22	25	8

3.4 Minimal polynomials

Let $\mathbf{F}_q$ be a subfield of $\mathbf{F}_r$. For an element α of $\mathbf{F}_r$, we are interested in nonzero polynomials $f(x) \in \mathbf{F}_q[x]$ of the least degree such that $f(\alpha) = 0$.

Definition 3.4.1 A *minimal polynomial* of an element $\alpha \in \mathbf{F}_{q^m}$ with respect to $\mathbf{F}_q$ is a nonzero monic polynomial $f(x)$ of the least degree in $\mathbf{F}_q[x]$ such that $f(\alpha) = 0$.

Example 3.4.2 Let α be a root of the polynomial $1 + x + x^2 \in \mathbf{F}_2[x]$. It is clear that the two linear polynomials x and $1 + x$ are not minimal polynomials of α. Therefore, $1 + x + x^2$ is a minimal polynomial of α.

Since $1 + (1 + \alpha) + (1 + \alpha)^2 = 1 + 1 + \alpha + 1 + \alpha^2 = 1 + \alpha + \alpha^2 = 0$ and $1 + \alpha$ is not a root of x or $1 + x$, $1 + x + x^2$ is also a minimal polynomial of $1 + \alpha$.

Theorem 3.4.3 (i) *The minimal polynomial of an element of $\mathbf{F}_{q^m}$ with respect to $\mathbf{F}_q$ exists and is unique. It is also irreducible over $\mathbf{F}_q$.*

(ii) *If a monic irreducible polynomial $M(x) \in \mathbf{F}_q[x]$ has $\alpha \in \mathbf{F}_{q^m}$ as a root, then it is the minimal polynomial of α with respect to $\mathbf{F}_q$.*

Proof. (i) Let α be an element of $\mathbf{F}_{q^m}$. As α is a root of $x^{q^m} - x$, we know the existence of a minimal polynomial of α.

Suppose that $M_1(x), M_2(x) \in \mathbf{F}_q[x]$ are both minimal polynomials of α. By the division algorithm, we have $M_1(x) = s(x)M_2(x) + r(x)$ for some polynomial $r(x)$ with $r(x) = 0$ or $\deg(r(x)) < \deg(M_2(x))$. Evaluating the polynomials at α, we obtain $0 = M_1(\alpha) = s(\alpha)M_2(\alpha) + r(\alpha) = r(\alpha)$. By the definition of minimal polynomials, this forces $r(x) = 0$; i.e., $M_2(x)|M_1(x)$. Similarly, we have $M_1(x)|M_2(x)$. Thus, we obtain $M_1(x) = M_2(x)$ since both are monic.

Let $M(x)$ be the minimal polynomial of α. Suppose that it is reducible. Then we have two monic polynomials $f(x) \in \mathbf{F}_q[x]$ and $g(x) \in \mathbf{F}_q[x]$ such that $\deg(f(x)) < \deg(M(x))$, $\deg(g(x)) < \deg(M(x))$ and $M(x) = f(x)g(x)$. Thus, we have $0 = M(\alpha) = f(\alpha)g(\alpha)$, which implies that $f(\alpha) = 0$ or $g(\alpha) = 0$. This contradicts the minimality of the degree of $M(x)$.

(ii) Let $f(x)$ be the minimal polynomial of α with respect to $\mathbf{F}_q$. By the division algorithm, there exist polynomials $h(x), e(x) \in \mathbf{F}_q[x]$ such that $M(x) = h(x)f(x) + e(x)$ and $\deg(e(x)) < \deg(f(x))$. Evaluating the polynomials at α, we obtain $0 = M(\alpha) = h(\alpha)f(\alpha) + e(\alpha) = e(\alpha)$. By the definition of the minimal polynomial, this forces $e(x) = 0$. This implies that $f(x)$ is the same as $M(x)$ since $M(x)$ is a monic irreducible polynomial and $f(x)$ cannot be a nonzero constant. This completes the proof. □

Example 3.4.4 Let $f(x) \in \mathbf{F}_q[x]$ be a monic irreducible polynomial of degree m. Let α be a root of $f(x)$. Then the minimal polynomial of $\alpha \in \mathbf{F}_{q^m}$ with respect to $\mathbf{F}_q$ is $f(x)$ itself. For instance, the minimal polynomial of a root of $2 + x + x^2 \in \mathbf{F}_3[x]$ is $2 + x + x^2$.

If we are given the minimal polynomial of a primitive element $\alpha \in \mathbf{F}_{q^m}$, we would like to find the minimal polynomial of α^i, for any i. In order to do so, we have to start with cyclotomic cosets.

Definition 3.4.5 Let n be co-prime to q. The *cyclotomic coset* of q (or q-*cyclotomic coset*) modulo n containing i is defined by

$$C_i = \{(i \cdot q^j \pmod n)) \in \mathbf{Z}_n : j = 0, 1, \ldots\}.$$

A subset $\{i_1, \ldots, i_t\}$ of $\mathbf{Z}_n$ is called a *complete set of representatives* of cyclotomic cosets of q modulo n if $C_{i_1}, \ldots, C_{i_t}$ are distinct and $\bigcup_{j=1}^{t} C_{i_j} = \mathbf{Z}_n$.

Remark 3.4.6 (i) It is easy to verify that two cyclotomic cosets are either equal or disjoint. Hence, the cyclotomic cosets partition $\mathbf{Z}_n$.

(ii) If $n = q^m - 1$ for some $m \geq 1$, each cyclotomic coset contains at most m elements, as $q^m \equiv 1 \pmod{q^m - 1}$.

(iii) It is easy to see that, in the case of $n = q^m - 1$ for some $m \geq 1$, $|C_i| = m$ if $\gcd(i, q^m - 1) = 1$.

Example 3.4.7 (i) Consider the cyclotomic cosets of 2 modulo 15:

$$C_0 = \{0\}, \qquad C_1 = \{1, 2, 4, 8\}, \qquad C_3 = \{3, 6, 9, 12\},$$
$$C_5 = \{5, 10\}, \quad C_7 = \{7, 11, 13, 14\}.$$

Thus, $C_1 = C_2 = C_4 = C_8$, and so on. The set $\{0, 1, 3, 5, 7\}$ is a complete set of representatives of cyclotomic cosets of 2 modulo 15. The set $\{0, 1, 6, 10, 7\}$ is also a complete set of representatives of cyclotomic cosets of 2 modulo 15.

(ii) Consider the cyclotomic cosets of 3 modulo 26:

$$C_0 = \{0\}, \qquad C_1 = \{1, 3, 9\}, \qquad C_2 = \{2, 6, 18\},$$
$$C_4 = \{4, 12, 10\}, \qquad C_5 = \{5, 15, 19\}, \quad C_7 = \{7, 21, 11\},$$
$$C_8 = \{8, 24, 20\}, \qquad C_{13} = \{13\}, \qquad C_{14} = \{14, 16, 22\},$$
$$C_{17} = \{17, 25, 23\}.$$

In this case, we have $C_1 = C_3 = C_9$, and so on. The set $\{0, 1, 2, 4, 5, 7, 8, 13, 14, 17\}$ is a complete set of representatives of cyclotomic cosets of 3 modulo 26.

We are now ready to determine the minimal polynomials for all the elements in a finite field.

Theorem 3.4.8 *Let α be a primitive element of $\mathbf{F}_{q^m}$. Then the minimal polynomial of α^i with respect to $\mathbf{F}_q$ is*

$$M^{(i)}(x) := \prod_{j \in C_i} (x - \alpha^j),$$

where C_i is the unique cyclotomic coset of q modulo $q^m - 1$ containing i.

Proof. *Step* 1: It is clear that α^i is a root of $M^{(i)}(x)$ as $i \in C_i$.

Step 2: Let $M^{(i)}(x) = a_0 + a_1 x + \cdots + a_r x^r$, where $a_k \in \mathbf{F}_{q^m}$ and $r = |C_i|$. Raising each coefficient to its qth power, we obtain

$$a_0^q + a_1^q x + \cdots + a_r^q x^r = \prod_{j \in C_i}(x - \alpha^{qj}) = \prod_{j \in C_{qi}}(x - \alpha^j)$$
$$= \prod_{j \in C_i}(x - \alpha^j) = M^{(i)}(x).$$

Note that we use the fact that $C_i = C_{qi}$ in the above formula. Hence, $a_k = a_k^q$ for all $0 \le k \le r$; i.e., a_k are elements of $\mathbf{F}_q$. This means that $M^{(i)}(x)$ is a polynomial over $\mathbf{F}_q$.

Step 3: Since α is a primitive element, we have $\alpha^j \ne \alpha^k$ for two distinct elements j, k of C_i. Hence, $M^{(i)}(x)$ has no multiple roots. Now let $f(x) \in \mathbf{F}_q[x]$ and $f(\alpha^i) = 0$. Put

$$f(x) = f_0 + f_1 x + \cdots + f_n x^n$$

for some $f_k \in \mathbf{F}_q$. Then, for any $j \in C_i$, there exists an integer l such that $j \equiv iq^l \pmod{q^m - 1}$. Hence,

$$
\begin{aligned}
f(\alpha^j) = f\left(\alpha^{iq^l}\right) &= f_0 + f_1 \alpha^{iq^l} + \cdots + f_n \alpha^{niq^l} \\
&= f_0^{q^l} + f_1^{q^l} \alpha^{iq^l} + \cdots + f_n^{q^l} \alpha^{niq^l} \\
&= (f_0 + f_1 \alpha^i + \cdots + f_n \alpha^{ni})^{q^l} \\
&= f(\alpha^i)^{q^l} = 0.
\end{aligned}
$$

This implies that $M^{(i)}(x)$ is a divisor of $f(x)$.

The above three steps show that $M^{(i)}(x)$ is the minimal polynomial of α^i. □

Remark 3.4.9 (i) The degree of the minimal polynomial of α^i is equal to the size of the cyclotomic coset containing i.

(ii) From Theorem 3.4.8, we know that α^i and α^k have the same minimal polynomial if and only if i, k are in the same cyclotomic coset.

Example 3.4.10 Let α be a root of $2 + x + x^2 \in \mathbf{F}_3[x]$; i.e.,

$$2 + \alpha + \alpha^2 = 0. \tag{3.4}$$

Then the minimal polynomial of α as well as α^3 is $2 + x + x^2$. The minimal polynomial of α^2 is

$$M^{(2)}(x) = \prod_{j \in C_2}(x - \alpha^j) = (x - \alpha^2)(x - \alpha^6) = \alpha^8 - (\alpha^2 + \alpha^6)x + x^2.$$

We know that $\alpha^8 = 1$ as $\alpha \in \mathbf{F}_9$. To find $M^{(2)}(x)$, we have to simplify $\alpha^2 + \alpha^6$. We make use of the relationship (3.4) to obtain

$$
\begin{aligned}
\alpha^2 + \alpha^6 &= (1 - \alpha) + (1 - \alpha)^3 = 2 - \alpha - \alpha^3 \\
&= 2 - \alpha - \alpha(1 - \alpha) = 2 - 2\alpha + \alpha^2 = 0.
\end{aligned}
$$

Hence, the minimal polynomial of α^2 is $1 + x^2$. In the same way, we may obtain the minimal polynomial $2 + 2x + x^2$ of α^5.

The following result will be useful when we study cyclic codes in Chapter 7.

Theorem 3.4.11 *Let n be a positive integer with $\gcd(q, n) = 1$. Suppose that m is a positive integer satisfying $n|(q^m - 1)$. Let α be a primitive element of $\mathbf{F}_{q^m}$ and let $M^{(j)}(x)$ be the minimal polynomial of α^j with respect to $\mathbf{F}_q$. Let $\{s_1, \ldots, s_t\}$ be a complete set of representatives of cyclotomic cosets of* *q modulo n. Then the polynomial $x^n - 1$ has the factorization into monic irreducible polynomials over $\mathbf{F}_q$:*

$$x^n - 1 = \prod_{i=1}^{t} M^{((q^m-1)s_i/n)}(x).$$

Proof. Put $r = (q^m - 1)/n$. Then α^r is a primitive nth root of unity, and hence all the roots of $x^n - 1$ are $1, \alpha^r, \alpha^{2r}, \ldots, \alpha^{(n-1)r}$. Thus, by the definition of the minimal polynomial, the polynomials $M^{(ir)}(x)$ are divisors of $x^n - 1$, for all $0 \le i \le n - 1$. It is clear that we have

$$x^n - 1 = \mathrm{lcm}(M^{(0)}(x), M^{(r)}(x), M^{(2r)}(x), \ldots, M^{((n-1)r)}(x)).$$

In order to determine the factorization of $x^n - 1$, it suffices to determine all the distinct polynomials among $M^{(0)}(x), M^{(r)}(x), M^{(2r)}(x), \ldots, M^{((n-1)r)}(x)$. By Remark 3.4.9, we know that $M^{(ir)}(x) = M^{(jr)}(x)$ if and only if ir and jr are in the same cyclotomic coset of q modulo $q^m - 1 = rn$; i.e., i and j are in the same cyclotomic coset of q modulo n. This implies that all the distinct polynomials among $M^{(0)}(x), M^{(r)}(x), M^{(2r)}(x), \ldots, M^{((n-1)r)}(x)$ are $M^{(s_1r)}(x), M^{(s_2r)}(x), \ldots, M^{(s_tr)}(x)$. The proof is completed. $\qquad\square$

The following result follows immediately from the above theorem.

Corollary 3.4.12 *Let n be a positive integer with $\gcd(q, n) = 1$. Then the number of monic irreducible factors of $x^n - 1$ over $\mathbf{F}_q$ is equal to the number of cyclotomic cosets of q modulo n.*

Example 3.4.13 (i) Consider the polynomial $x^{13} - 1$ over $\mathbf{F}_3$. It is easy to check that $\{0, 1, 2, 4, 7\}$ is a complete set of representatives of cyclotomic cosets of 3 modulo 13. Since 13 is a divisor of $3^3 - 1$, we consider the field $\mathbf{F}_{27}$. Let α be a root of $1 + 2x + x^3$. By Example 3.3.12, α is a primitive element of $\mathbf{F}_{27}$. By Example 3.4.7(ii), we know all the cyclotomic cosets of 3 modulo 26 containing multiples of 2. Hence, we obtain

$$M^{(0)}(x) = 2 + x,$$
$$M^{(2)}(x) = \prod_{j \in C_2}(x - \alpha^j) = (x - \alpha^2)(x - \alpha^6)(x - \alpha^{18}) = 2 + x + x^2 + x^3,$$
$$M^{(4)}(x) = \prod_{j \in C_4}(x - \alpha^j) = (x - \alpha^4)(x - \alpha^{12})(x - \alpha^{10}) = 2 + x^2 + x^3,$$
$$M^{(8)}(x) = \prod_{j \in C_8}(x - \alpha^j) = (x - \alpha^8)(x - \alpha^{20})(x - \alpha^{24})$$
$$= 2 + 2x + 2x^2 + x^3,$$

$$M^{(14)}(x) = \prod_{j \in C_{14}} (x - \alpha^j) = (x - \alpha^{14})(x - \alpha^{16})(x - \alpha^{22}) = 2 + 2x + x^3.$$

By Theorem 3.4.11, we obtain the factorization of $x^{13} - 1$ over $\mathbf{F}_3$ into monic irreducible polynomials:

$$\begin{aligned}
x^{13} - 1 &= M^{(0)}(x)M^{(2)}(x)M^{(4)}(x)M^{(8)}(x)M^{(14)}(x) \\
&= (2 + x)(2 + x + x^2 + x^3)(2 + x^2 + x^3) \\
&\quad \times (2 + 2x + 2x^2 + x^3)(2 + 2x + x^3).
\end{aligned}$$

(ii) Consider the polynomial $x^{21} - 1$ over $\mathbf{F}_2$. It is easy to check that $\{0, 1, 3, 5, 7, 9\}$ is a complete set of representatives of cyclotomic cosets of 2 modulo 21. Since 21 is a divisor of $2^6 - 1$, we consider the field $\mathbf{F}_{64}$. Let α be a root of $1 + x + x^6$. It can be verified that α is a primitive element of $\mathbf{F}_{64}$ (check that $\alpha^3 \neq 1$, $\alpha^7 \neq 1$, $\alpha^9 \neq 1$ and $\alpha^{21} \neq 1$). We list the cyclotomic cosets of 2 modulo 63 containing multiples of 3:

$$\begin{aligned}
C_0 &= \{0\}, & C_3 &= \{3, 6, 12, 24, 48, 33\}, \\
C_9 &= \{9, 18, 36\}, & C_{15} &= \{15, 30, 60, 57, 51, 39\}, \\
C_{21} &= \{21, 42\}, & C_{27} &= \{27, 54, 45\}.
\end{aligned}$$

Hence, we obtain

$$\begin{aligned}
M^{(0)}(x) &= 1 + x, \\
M^{(3)}(x) &= \prod_{j \in C_3} (x - \alpha^j) = 1 + x + x^2 + x^4 + x^6, \\
M^{(9)}(x) &= \prod_{j \in C_9} (x - \alpha^j) = 1 + x^2 + x^3, \\
M^{(15)}(x) &= \prod_{j \in C_{15}} (x - \alpha^j) = 1 + x^2 + x^4 + x^5 + x^6, \\
M^{(21)}(x) &= \prod_{j \in C_{21}} (x - \alpha^j) = 1 + x + x^2, \\
M^{(27)}(x) &= \prod_{j \in C_{27}} (x - \alpha^j) = 1 + x + x^3.
\end{aligned}$$

By Theorem 3.4.11, we obtain the factorization of $x^{21} - 1$ over $\mathbf{F}_2$ into monic irreducible polynomials:

$$\begin{aligned}
x^{21} - 1 &= M^{(0)}(x)M^{(3)}(x)M^{(9)}(x)M^{(15)}(x)M^{(21)}(x)M^{(27)}(x) \\
&= (1 + x)(1 + x + x^2 + x^4 + x^6)(1 + x^2 + x^3) \\
&\quad \times (1 + x^2 + x^4 + x^5 + x^6)(1 + x + x^2)(1 + x + x^3).
\end{aligned}$$

Exercises

3.1 Show that the remainder of every square integer divided by 4 is either 0 or 1. Hence, show that there do not exist integers x and y such that $x^2 + y^2 = 40\,403$.

3.2 Construct the addition and multiplication tables for the rings $\mathbf{Z}_5$ and $\mathbf{Z}_8$.

3.3 Find the multiplicative inverse of each of the following elements:
   (a) $2, 5$ and 8 in $\mathbf{Z}_{11}$,
   (b) $4, 7$ and 11 in $\mathbf{Z}_{17}$.

3.4 Let p be a prime.
   (i) Show that $\binom{p}{j} \equiv 0 \pmod{p}$ for any $1 \le j \le p - 1$.
   (ii) Show that $\binom{p-1}{j} \equiv (-1)^j \pmod{p}$ for any $1 \le j \le p - 1$.
   (iii) Show that, for any two elements α, β in a field of characteristic p, we have

$$(\alpha + \beta)^{p^k} = \alpha^{p^k} + \beta^{p^k}$$

   for any $k \ge 0$.

3.5 (i) Verify that $f(x) = x^5 - 1 \in \mathbf{F}_{31}[x]$ can be written as the product $(x^2 - 3x + 2)(x^3 + 3x^2 + 7x + 15)$.
   (ii) Determine the remainder of $f(x)$ divided by $x^2 - 3x + 2$.
   (iii) Compute the remainders of $f(x)$ divided by x^5, x^7 and $x^4 + 5x^3$, respectively.

3.6 Verify that the following polynomials are irreducible:
   (a) $1 + x + x^2 + x^3 + x^4$, $1 + x + x^4$ and $1 + x^3 + x^4$ over $\mathbf{F}_2$;
   (b) $1 + x^2$, $2 + x + x^2$ and $2 + 2x + x^2$ over $\mathbf{F}_3$.

3.7 Every quadratic or cubic polynomial is either irreducible or has a linear factor.
   (a) Find the number of monic irreducible quadratic polynomials over $\mathbf{F}_q$.
   (b) Find the number of monic irreducible cubic polynomials over $\mathbf{F}_q$.
   (c) Determine all the irreducible quadratic and cubic polynomials over $\mathbf{F}_2$.
   (d) Determine all the monic irreducible quadratic polynomials over $\mathbf{F}_3$.

3.8 Let $f(x) = (2 + 2x^2)(2 + x^2 + x^3)^2(-1 + x^4) \in \mathbf{F}_3[x]$ and $g(x) = (1 + x^2)(-2 + 2x^2)(2 + x^2 + x^3) \in \mathbf{F}_3[x]$. Determine $\gcd(f(x), g(x))$ and $\operatorname{lcm}(f(x), g(x))$.

3.9 Find two polynomials $u(x)$ and $v(x) \in \mathbf{F}_2[x]$ such that $\deg(u(x)) < 4$, $\deg(v(x)) < 3$ and

$$u(x)(1 + x + x^3) + v(x)(1 + x + x^2 + x^3 + x^4) = 1.$$

3.10 Construct both the addition and multiplication tables for the ring $\mathbf{F}_3[x]/(x^2 + 2)$.

3.11 (a) Find the order of the elements 2, 7, 10 and 12 in $\mathbf{F}_{17}$.

   (b) Find the order of the elements α, α^3, $\alpha + 1$ and $\alpha^3 + 1$ in $\mathbf{F}_{16}$, where α is a root of $1 + x + x^4$.

3.12 (i) Let α be a primitive element of $\mathbf{F}_q$. Show that α^i is also a primitive element if and only if $\gcd(i, q - 1) = 1$.

   (ii) Determine the number of primitive elements in the following fields: $\mathbf{F}_9$, $\mathbf{F}_{19}$, $\mathbf{F}_{25}$ and $\mathbf{F}_{32}$.

3.13 Determine all the primitive elements of the following fields: $\mathbf{F}_7$, $\mathbf{F}_8$ and $\mathbf{F}_9$.

3.14 Show that all the nonzero elements, except the identity 1, in $\mathbf{F}_{128}$ are primitive elements.

3.15 Show that any root of $1 + x + x^6 \in \mathbf{F}_2[x]$ is a primitive element of $\mathbf{F}_{64}$.

3.16 Consider the field with 16 elements constructed using the irreducible polynomial $f(x) = 1 + x^3 + x^4$ over $\mathbf{F}_2$.

   (i) Let α be a root of $f(x)$. Show that α is a primitive element of $\mathbf{F}_{16}$. Represent each element both as a polynomial and as a power of α.

   (ii) Construct a Zech's log table for the field.

3.17 Find a primitive element and construct a Zech's log table for each of the following finite fields:

   (a) $\mathbf{F}_{3^2}$,   (b) $\mathbf{F}_{2^5}$,   (c) $\mathbf{F}_{5^2}$.

3.18 Show that each monic irreducible polynomial of $\mathbf{F}_q[x]$ of degree m is the minimal polynomial of some element of $\mathbf{F}_{q^m}$ with respect to $\mathbf{F}_q$.

3.19 Let α be a root of $1 + x^3 + x^4 \in \mathbf{F}_2[x]$.

   (i) List all the cyclotomic cosets of 2 modulo 15.

   (ii) Find the minimal polynomial of $\alpha^i \in \mathbf{F}_{16}$, for all $1 \le i \le 14$.

   (iii) Using Exercise 3.18, find all the irreducible polynomials of degree 4 over $\mathbf{F}_2$.

3.20 (i) Find all the cyclotomic cosets of 2 modulo 31.

   (ii) Find the minimal polynomials of α, α^4 and α^5, where α is a root of $1 + x^2 + x^5 \in \mathbf{F}_2[x]$.

3.21 Based on the cyclotomic cosets of 3 modulo 26, find all the monic irreducible polynomials of degree 3 over $\mathbf{F}_3$.

3.22 (i) Prove that, if k is a positive divisor of m, then $\mathbf{F}_{p^m}$ contains a unique subfield with p^k elements.

   (ii) Determine all the subfields in (a) $\mathbf{F}_{2^{12}}$, (b) $\mathbf{F}_{2^{18}}$.

3.23 Factorize the following polynomials:

   (a) $x^7 - 1$ over $\mathbf{F}_2$;   (b) $x^{15} - 1$ over $\mathbf{F}_2$;

   (c) $x^{31} - 1$ over $\mathbf{F}_2$;   (d) $x^8 - 1$ over $\mathbf{F}_3$;

   (e) $x^{12} - 1$ over $\mathbf{F}_5$;   (f) $x^{24} - 1$ over $\mathbf{F}_7$.

3.24 Show that, for any given element c of $\mathbf{F}_q$, there exist two elements a and b of $\mathbf{F}_q$ such that $a^2 + b^2 = c$.

3.25 For a nonzero element b of $\mathbf{F}_p$, where p is a prime, prove that the trinomial $x^p - x - b$ is irreducible in $\mathbf{F}_{p^n}[x]$ if and only if n is not divisible by p.

3.26 (Lagrange interpolation formula.) For $n \geq 1$, let $\alpha_1, \ldots, \alpha_n$ be n distinct elements of $\mathbf{F}_q$, and let $\beta_1, \ldots, \beta_n$ be n arbitrary elements of $\mathbf{F}_q$. Show that there exists exactly one polynomial $f(x) \in \mathbf{F}_q[x]$ of degree $\leq n - 1$ such that $f(\alpha_i) = \beta_i$ for $i = 1, \ldots, n$. Furthermore, show that this polynomial is given by

$$f(x) = \sum_{i=1}^{n} \frac{\beta_i}{g'(\alpha_i)} \prod_{\substack{k=1 \\ k \neq i}}^{n} (x - \alpha_k),$$

where $g'(x)$ denotes the derivative of $g(x) := \prod_{k=1}^{n}(x - \alpha_k)$.

3.27 (i) Show that, for every integer $n \geq 1$, the product of all monic irreducible polynomials over $\mathbf{F}_q$ whose degrees divide n is equal to $x^{q^n} - x$.

(ii) Let $I_q(d)$ denote the number of monic irreducible polynomials of degree d in $\mathbf{F}_q[x]$. Show that

$$q^n = \sum_{d \mid n} d I_q(d) \quad \text{for all } n \in \mathbf{N},$$

where the sum is extended over all positive divisors d of n.

3.28 The *Möbius function* on the set $\mathbf{N}$ of positive integers is defined by

$$\mu(n) = \begin{cases} 1 & \text{if } n = 1 \\ (-1)^k & \text{if } n \text{ is a product of } k \text{ distinct primes} \\ 0 & \text{if } n \text{ is divisible by the square of a prime.} \end{cases}$$

(i) Let h and H be two functions from $\mathbf{N}$ to $\mathbf{Z}$. Show that

$$H(n) = \sum_{d \mid n} h(d)$$

for all $n \in \mathbf{N}$ if and only if

$$h(n) = \sum_{d \mid n} \mu(d) H\left(\frac{n}{d}\right)$$

for all $n \in \mathbf{N}$.

(ii) Show that the number $I_q(n)$ of monic irreducible polynomials over $\mathbf{F}_q$ of degree n is given by

$$I_q(n) = \frac{1}{n} \sum_{d \mid n} \mu(d) q^{n/d}.$$

4   Linear codes

A linear code of length n over the finite field $\mathbf{F}_q$ is simply a subspace of the vector space $\mathbf{F}_q^n$. Since linear codes are vector spaces, their algebraic structures often make them easier to describe and use than nonlinear codes. In most of this book, we focus our attention on linear codes over finite fields.

4.1   Vector spaces over finite fields

We recall some definitions and facts about vector spaces over finite fields. While the proofs of most of the facts stated in this section are omitted, it should be noted that many of them are practically identical to those in the case of vector spaces over $\mathbf{R}$ or $\mathbf{C}$.

Definition 4.1.1 Let $\mathbf{F}_q$ be the finite field of order q. A nonempty set V, together with some (vector) addition $+$ and scalar multiplication by elements of $\mathbf{F}_q$, is a *vector space* (or *linear space*) over $\mathbf{F}_q$ if it satisfies all of the following conditions. For all $\mathbf{u}, \mathbf{v}, \mathbf{w} \in V$ and for all $\lambda, \mu \in \mathbf{F}_q$:

  (i) $\mathbf{u} + \mathbf{v} \in V$;

  (ii) $(\mathbf{u} + \mathbf{v}) + \mathbf{w} = \mathbf{u} + (\mathbf{v} + \mathbf{w})$;

  (iii) there is an element $\mathbf{0} \in V$ with the property $\mathbf{0} + \mathbf{v} = \mathbf{v} = \mathbf{v} + \mathbf{0}$ for all $\mathbf{v} \in V$;

  (iv) for each $\mathbf{u} \in V$ there is an element of V, called $-\mathbf{u}$, such that $\mathbf{u} + (-\mathbf{u}) = \mathbf{0} = (-\mathbf{u}) + \mathbf{u}$;

  (v) $\mathbf{u} + \mathbf{v} = \mathbf{v} + \mathbf{u}$;

  (vi) $\lambda \mathbf{v} \in V$;

  (vii) $\lambda(\mathbf{u} + \mathbf{v}) = \lambda \mathbf{u} + \lambda \mathbf{v}, (\lambda + \mu)\mathbf{u} = \lambda \mathbf{u} + \mu \mathbf{u}$;

  (viii) $(\lambda \mu)\mathbf{u} = \lambda(\mu \mathbf{u})$;

  (ix) if 1 is the multiplicative identity of $\mathbf{F}_q$, then $1\mathbf{u} = \mathbf{u}$.

Let $\mathbf{F}_q^n$ be the set of all vectors of length n with entries in $\mathbf{F}_q$:

$$\mathbf{F}_q^n = \{(v_1, v_2, \ldots, v_n) : v_i \in \mathbf{F}_q\}.$$

We define the *vector addition* for $\mathbf{F}_q^n$ componentwise, using the addition defined on $\mathbf{F}_q$; i.e., if

$$\mathbf{v} = (v_1, \ldots, v_n) \in \mathbf{F}_q^n \text{ and } \mathbf{w} = (w_1, \ldots, w_n) \in \mathbf{F}_q^n,$$

then

$$\mathbf{v} + \mathbf{w} = (v_1 + w_1, \ldots, v_n + w_n) \in \mathbf{F}_q^n.$$

We also define the *scalar multiplication* for $\mathbf{F}_q^n$ componentwise; i.e., if

$$\mathbf{v} = (v_1, \ldots, v_n) \in \mathbf{F}_q^n \text{ and } \lambda \in \mathbf{F}_q,$$

then

$$\lambda\mathbf{v} = (\lambda v_1, \ldots, \lambda v_n) \in \mathbf{F}_q^n.$$

Let $\mathbf{0}$ denote the zero vector $(0, 0, \ldots, 0) \in \mathbf{F}_q^n$.

Example 4.1.2 It is easy to verify that the following are vector spaces over $\mathbf{F}_q$:

  (i) (any q) $C_1 = \mathbf{F}_q^n$ and $C_2 = \{\mathbf{0}\}$;
  (ii) (any q) $C_3 = \{(\lambda, \ldots, \lambda) : \lambda \in \mathbf{F}_q\}$;
 (iii) ($q = 2$) $C_4 = \{(0, 0, 0, 0), (1, 0, 1, 0), (0, 1, 0, 1), (1, 1, 1, 1)\}$;
 (iv) ($q = 3$) $C_5 = \{(0, 0, 0), (0, 1, 2), (0, 2, 1)\}$.

Remark 4.1.3 When no confusion arises, it is sometimes convenient to write a vector $(v_1, v_2, \ldots, v_n)$ simply as $v_1 v_2 \cdots v_n$.

Definition 4.1.4 A nonempty subset C of a vector space V is a *subspace* of V if it is itself a vector space with the same vector addition and scalar multiplication as V.

Example 4.1.5 Using the same notation as in Example 4.1.2, it is easy to see that:

  (i) (any q) $C_2 = \{\mathbf{0}\}$ is a subspace of both C_3 and $C_1 = \mathbf{F}_q^n$, and C_3 is a subspace of $C_1 = \mathbf{F}_q^n$;
  (ii) ($q = 2$) C_4 is a subspace of $\mathbf{F}_2^4$;
 (iii) ($q = 3$) C_5 is a subspace of $\mathbf{F}_3^3$.

Proposition 4.1.6 *A nonempty subset C of a vector space V over* $\mathbf{F}_q$ *is a subspace if and only if the following condition is satisfied:*

$$\text{if } \mathbf{x}, \mathbf{y} \in C \text{ and } \lambda, \mu \in \mathbf{F}_q, \text{ then } \lambda\mathbf{x} + \mu\mathbf{y} \in C.$$

We leave the proof of Proposition 4.1.6 as an exercise (see Exercise 4.1). Note that, when $q = 2$, a necessary and sufficient condition for a nonempty subset C of a vector space V over $\mathbf{F}_2$ to be a subspace is: if $\mathbf{x}, \mathbf{y} \in C$, then $\mathbf{x} + \mathbf{y} \in C$.

Definition 4.1.7 Let V be a vector space over $\mathbf{F}_q$. A *linear combination* of $\mathbf{v}_1, \ldots, \mathbf{v}_r \in V$ is a vector of the form $\lambda_1\mathbf{v}_1 + \cdots + \lambda_r\mathbf{v}_r$, where $\lambda_1, \ldots, \lambda_r \in \mathbf{F}_q$ are some scalars.

Definition 4.1.8 Let V be a vector space over $\mathbf{F}_q$. A set of vectors $\{\mathbf{v}_1, \ldots, \mathbf{v}_r\}$ in V is *linearly independent* if

$$\lambda_1\mathbf{v}_1 + \cdots + \lambda_r\mathbf{v}_r = \mathbf{0} \Rightarrow \lambda_1 = \cdots = \lambda_r = 0.$$

The set is *linearly dependent* if it is not linearly independent; i.e., if there are $\lambda_1, \ldots, \lambda_r \in \mathbf{F}_q$, not all zero (but maybe some are!), such that $\lambda_1\mathbf{v}_1 + \cdots + \lambda_r\mathbf{v}_r = \mathbf{0}$.

Example 4.1.9 (i) Any set S which contains $\mathbf{0}$ is linearly dependent.

(ii) For any $\mathbf{F}_q$, $\{(0, 0, 0, 1), (0, 0, 1, 0), (0, 1, 0, 0)\}$ is linearly independent.

(iii) For any $\mathbf{F}_q$, $\{(0, 0, 0, 1), (1, 0, 0, 0), (1, 0, 0, 1)\}$ is linearly dependent.

Definition 4.1.10 Let V be a vector space over $\mathbf{F}_q$ and let $S = \{\mathbf{v}_1, \mathbf{v}_2, \ldots, \mathbf{v}_k\}$ be a nonempty subset of V. The *(linear) span* of S is defined as

$$<S> = \{\lambda_1\mathbf{v}_1 + \cdots + \lambda_k\mathbf{v}_k : \lambda_i \in \mathbf{F}_q\}.$$

If $S = \emptyset$, we define $<S> = \{\mathbf{0}\}$. It is easy to verify that $<S>$ is a subspace of V, called the subspace *generated* (or *spanned*) by S. Given a subspace C of V, a subset S of C is called a *generating set* (or *spanning set*) of C if $C = <S>$.

Remark 4.1.11 If S is already a subspace of V, then $<S> = S$.

Example 4.1.12 (i) If $q = 2$ and $S = \{0001, 0010, 0100\}$, then

$$<S> = \{0000, 0001, 0010, 0100, 0011, 0101, 0110, 0111\}.$$

(ii) If $q = 2$ and $S = \{0001, 1000, 1001\}$, then

$$<S> = \{0000, 0001, 1000, 1001\}.$$

(iii) If $q = 3$ and $S = \{0001, 1000, 1001\}$, then

$$<S> = \{0000, 0001, 0002, 1000, 2000, 1001, 1002, 2001, 2002\}.$$

Definition 4.1.13 Let V be a vector space over $\mathbf{F}_q$. A nonempty subset $B = \{\mathbf{v}_1, \mathbf{v}_2, \dots, \mathbf{v}_k\}$ of V is called a *basis* for V if $V = <B>$ and B is linearly independent.

Remark 4.1.14 (i) If $B = \{\mathbf{v}_1, \dots, \mathbf{v}_k\}$ is a basis of V, then any vector $\mathbf{v} \in V$ can be expressed as a unique linear combination of vectors in B; i.e., there exist unique $\lambda_1, \lambda_2, \dots, \lambda_k \in \mathbf{F}_q$ such that

$$\mathbf{v} = \lambda_1 \mathbf{v}_1 + \lambda_2 \mathbf{v}_2 + \cdots + \lambda_k \mathbf{v}_k.$$

(ii) A vector space V over a finite field $\mathbf{F}_q$ can have many bases; but all bases contain the same number of elements. This number is called the *dimension* of V over $\mathbf{F}_q$, denoted by $\dim(V)$. In the case where V can be regarded as a vector space over more than one field, the notation $\dim_{\mathbf{F}_q}(V)$ may be used to avoid confusion.

Theorem 4.1.15 *Let V be a vector space over $\mathbf{F}_q$. If $\dim(V) = k$, then*

(i) *V has q^k elements;*
(ii) *V has $\frac{1}{k!} \prod_{i=0}^{k-1} (q^k - q^i)$ different bases.*

Proof. (i) If $\{\mathbf{v}_1, \dots, \mathbf{v}_k\}$ is a basis for V, then

$$V = \{\lambda_1 \mathbf{v}_1 + \cdots + \lambda_k \mathbf{v}_k : \lambda_1, \dots, \lambda_k \in \mathbf{F}_q\}.$$

Since $|\mathbf{F}_q| = q$, there are exactly q choices for each of $\lambda_1, \dots, \lambda_k$; hence, V has exactly q^k elements.

(ii) Let $B = \{\mathbf{v}_1, \dots, \mathbf{v}_k\}$ denote a basis for V. Since $\mathbf{v}_1 \neq \mathbf{0}$, there are $q^k - 1$ choices for $\mathbf{v}_1$. For B to be a basis, the condition $\mathbf{v}_2 \notin <\mathbf{v}_1>$ is needed, so there are $q^k - q$ choices for $\mathbf{v}_2$. Arguing in this manner, for every i such that $k \geq i \geq 2$, we need $\mathbf{v}_i \notin <\mathbf{v}_1, \dots, \mathbf{v}_{i-1}>$, so there are $q^k - q^{i-1}$ choices for $\mathbf{v}_i$. Hence, there are $\prod_{i=0}^{k-1}(q^k - q^i)$ distinct *ordered* k-tuples $(\mathbf{v}_1, \dots, \mathbf{v}_k)$. However, since the order of $\mathbf{v}_1, \dots, \mathbf{v}_k$ is irrelevant for a basis, the number of distinct bases for V is $\frac{1}{k!} \prod_{i=0}^{k-1} (q^k - q^i)$. $\square$

Example 4.1.16 Let $q = 2$, $S = \{0001, 0010, 0100\}$ and $V = <S>$, then

$$V = \{0000, 0001, 0010, 0100, 0011, 0101, 0110, 0111\}.$$

Note that S is linearly independent, so $\dim(V) = 3$. We see that $|V| = 8 = 2^3$. By Theorem 4.1.15, the number of different bases for V is given by

$$\frac{1}{k!} \prod_{i=0}^{k-1}(2^k - 2^i) = \frac{1}{3!}(2^3 - 1)(2^3 - 2)(2^3 - 2^2) = 28.$$

Definition 4.1.17 Let $\mathbf{v} = (v_1, v_2, \ldots, v_n)$, $\mathbf{w} = (w_1, w_2, \ldots, w_n) \in \mathbf{F}_q^n$.

(i) The *scalar product* (also known as the *dot product* or the *Euclidean inner product*) of $\mathbf{v}$ and $\mathbf{w}$ is defined as

$$\mathbf{v} \cdot \mathbf{w} = v_1 w_1 + \cdots + v_n w_n \in \mathbf{F}_q.$$

(ii) The two vectors $\mathbf{v}$ and $\mathbf{w}$ are said to be *orthogonal* if $\mathbf{v} \cdot \mathbf{w} = 0$.

(iii) Let S be a nonempty subset of $\mathbf{F}_q^n$. The *orthogonal complement* $S^\perp$ of S is defined to be

$$S^\perp = \{\mathbf{v} \in \mathbf{F}_q^n : \mathbf{v} \cdot \mathbf{s} = 0 \text{ for all } \mathbf{s} \in S\}.$$

If $S = \emptyset$, then we define $S^\perp = \mathbf{F}_q^n$.

Remark 4.1.18 (i) It is easy to verify that $S^\perp$ is always a subspace of the vector space $\mathbf{F}_q^n$ for any subset S of $\mathbf{F}_q^n$, and that $<S>^\perp = S^\perp$.

(ii) The scalar product is an example of an *inner product* on $\mathbf{F}_q^n$. An inner product on $\mathbf{F}_q^n$ is a pairing $\langle,\rangle : \mathbf{F}_q^n \times \mathbf{F}_q^n \to \mathbf{F}_q$ satisfying the following conditions: for all $\mathbf{u}, \mathbf{v}, \mathbf{w} \in \mathbf{F}_q^n$,

(a) $\langle \mathbf{u} + \mathbf{v}, \mathbf{w} \rangle = \langle \mathbf{u}, \mathbf{w} \rangle + \langle \mathbf{v}, \mathbf{w} \rangle$;
(b) $\langle \mathbf{u}, \mathbf{v} + \mathbf{w} \rangle = \langle \mathbf{u}, \mathbf{v} \rangle + \langle \mathbf{u}, \mathbf{w} \rangle$;
(c) $\langle \mathbf{u}, \mathbf{v} \rangle = 0$ for all $\mathbf{u} \in \mathbf{F}_q^n$ if and only if $\mathbf{v} = \mathbf{0}$;
(d) $\langle \mathbf{u}, \mathbf{v} \rangle = 0$ for all $\mathbf{v} \in \mathbf{F}_q^n$ if and only if $\mathbf{u} = \mathbf{0}$.

The scalar product in Definition 4.1.17 is often called the Euclidean inner product. Some other inner products, such as the Hermitian inner product and symplectic inner product, are also used in coding theory (see Exercises 4.9–4.13). Throughout this book, unless it is otherwise specified, the inner product used is always assumed to be the scalar product, i.e., the Euclidean inner product.

Example 4.1.19 (i) Let $q = 2$ and let $n = 4$. If $\mathbf{u} = (1, 1, 1, 1)$, $\mathbf{v} = (1, 1, 1, 0)$, $\mathbf{w} = (1, 0, 0, 1)$, then

$$\mathbf{u} \cdot \mathbf{v} = 1 \cdot 1 + 1 \cdot 1 + 1 \cdot 1 + 1 \cdot 0 = 1,$$
$$\mathbf{u} \cdot \mathbf{w} = 1 \cdot 1 + 1 \cdot 0 + 1 \cdot 0 + 1 \cdot 1 = 0,$$
$$\mathbf{v} \cdot \mathbf{w} = 1 \cdot 1 + 1 \cdot 0 + 1 \cdot 0 + 0 \cdot 1 = 1.$$

Hence, $\mathbf{u}$ and $\mathbf{w}$ are orthogonal.

(ii) Let $q = 2$ and let $S = \{0100, 0101\}$. To find $S^\perp$, let $\mathbf{v} = (v_1, v_2, v_3, v_4) \in S^\perp$. Then

$$\mathbf{v} \cdot (0, 1, 0, 0) = 0 \Rightarrow v_2 = 0,$$
$$\mathbf{v} \cdot (0, 1, 0, 1) = 0 \Rightarrow v_2 + v_4 = 0.$$

Hence, we have $v_2 = v_4 = 0$. Since v_1 and v_3 can be either 0 or 1, we can conclude that

$$S^\perp = \{0000, 0010, 1000, 1010\}.$$

Theorem 4.1.20 *Let S be a subset of $\mathbf{F}_q^n$, then we have*

$$\dim(<S>) + \dim(S^\perp) = n.$$

Proof. Theorem 4.1.20 is obviously true when $<S> = \{\mathbf{0}\}$.

Now let $\dim(<S>) = k \geq 1$ and suppose $\{\mathbf{v}_1, \ldots, \mathbf{v}_k\}$ is a basis of $<S>$. We need to show that $\dim(S^\perp) = \dim(<S>^\perp) = n - k$.

Note that $\mathbf{x} \in S^\perp$ if and only if

$$\mathbf{v}_1 \cdot \mathbf{x} = \cdots = \mathbf{v}_k \cdot \mathbf{x} = 0,$$

which is equivalent to saying that $\mathbf{x}$ satisfies $A\mathbf{x}^T = \mathbf{0}$, where A is the $k \times n$ matrix whose ith row is $\mathbf{v}_i$.

The rows of A are linearly independent, so $A\mathbf{x}^T = \mathbf{0}$ is a linear system of k linearly independent equations in n variables. From linear algebra, it is known that such a system admits a solution space of dimension $n - k$. □

Example 4.1.21 Let $q = 2$, $n = 4$ and $S = \{0100, 0101\}$. Then

$$<S> = \{0000, 0100, 0001, 0101\}.$$

Note that S is linearly independent, so $\dim(<S>) = 2$. We have computed that (Example 4.1.19)

$$S^\perp = \{0000, 0010, 1000, 1010\}.$$

Note that $\{0010, 1000\}$ is a basis for $S^\perp$, so $\dim(S^\perp) = 2$. Hence, we have verified that

$$\dim(<S>) + \dim(S^\perp) = 2 + 2 = 4 = n.$$

4.2   Linear codes

We are now ready to introduce linear codes and discuss some of their elementary properties.

Definition 4.2.1 A *linear code* C of length n over $\mathbf{F}_q$ is a subspace of $\mathbf{F}_q^n$.

Example 4.2.2 The following are linear codes:

(i) $C = \{(\lambda, \lambda, \ldots, \lambda) : \lambda \in \mathbf{F}_q\}$. This code is often called a *repetition code* (refer also to Example 1.0.3).
(ii) $(q = 2)\, C = \{000, 001, 010, 011\}$.
(iii) $(q = 3)\, C = \{0000, 1100, 2200, 0001, 0002, 1101, 1102, 2201, 2202\}$.
(iv) $(q = 2)\, C = \{000, 001, 010, 011, 100, 101, 110, 111\}$.

Definition 4.2.3 Let C be a linear code in $\mathbf{F}_q^n$.
  (i) The *dual code* of C is $C^\perp$, the orthogonal complement of the subspace C of $\mathbf{F}_q^n$.
  (ii) The *dimension* of the linear code C is the dimension of C as a vector space over $\mathbf{F}_q$, i.e., $\dim(C)$.

Theorem 4.2.4 *Let C be a linear code of length n over $\mathbf{F}_q$. Then,*

(i) $|C| = q^{\dim(C)}$, *i.e.,* $\dim(C) = \log_q |C|$;
(ii) $C^\perp$ *is a linear code and* $\dim(C) + \dim(C^\perp) = n$;
(iii) $(C^\perp)^\perp = C$.

Proof. (i) follows from Theorem 4.1.15(i).
  (ii) follows immediately from Remark 4.1.18(i) and Theorem 4.1.20 with $C = S$.
  Using the equality in (ii) and a similar equality with C replaced by $C^\perp$, we obtain $\dim(C) = \dim((C^\perp)^\perp)$. To prove (iii), it therefore suffices to show that $C \subseteq (C^\perp)^\perp$.
  Let $\mathbf{c} \in C$. To show that $\mathbf{c} \in (C^\perp)^\perp$, we need to show that $\mathbf{c} \cdot \mathbf{x} = 0$ for all $\mathbf{x} \in C^\perp$. Since $\mathbf{c} \in C$ and $\mathbf{x} \in C^\perp$, by the definition of $C^\perp$, it follows that $\mathbf{c} \cdot \mathbf{x} = 0$. Hence, (iii) is proved.   □

Example 4.2.5 (i) ($q = 2$) Let $C = \{0000, 1010, 0101, 1111\}$, so $\dim(C) = \log_2 |C| = \log_2 4 = 2$. It is easy to see that $C^{\perp} = \{0000, 1010, 0101, 1111\} = C$, so $\dim(C^{\perp}) = 2$. In particular, Theorem 4.2.4(ii) and (iii) are verified.

(ii) ($q = 3$) Let $C = \{000, 001, 002, 010, 020, 011, 012, 021, 022\}$, so $\dim(C) = \log_3 |C| = \log_3 9 = 2$. One checks readily that $C^{\perp} = \{000, 100, 200\}$, so $\dim(C^{\perp}) = 1$.

Remark 4.2.6 A linear code C of length n and dimension k over $\mathbf{F}_q$ is often called a q-ary $[n, k]$-code or, if q is clear from the context, an $[n, k]$-code. It is also an (n, q^k)-linear code. If the distance d of C is known, it is also sometimes referred to as an $[n, k, d]$-linear code.

Definition 4.2.7 Let C be a linear code.
   (i) C is *self-orthogonal* if $C \subseteq C^{\perp}$.
   (ii) C is *self-dual* if $C = C^{\perp}$.

Proposition 4.2.8 *The dimension of a self-orthogonal code of length n must be $\leq n/2$, and the dimension of a self-dual code of length n is $n/2$.*

Proof. This proposition is an immediate consequence of Theorem 4.2.4(ii) and the definitions of self-orthogonal and self-dual codes. $\square$

Example 4.2.9 The code in Example 4.2.5(i) is self-dual.

4.3   Hamming weight

Recall that the Hamming distance $d(\mathbf{x}, \mathbf{y})$ between two words $\mathbf{x}, \mathbf{y} \in \mathbf{F}_q^n$ was defined in Chapter 2.

Definition 4.3.1 Let $\mathbf{x}$ be a word in $\mathbf{F}_q^n$. The (*Hamming*) *weight* of $\mathbf{x}$, denoted by $\mathrm{wt}(\mathbf{x})$, is defined to be the number of nonzero coordinates in $\mathbf{x}$; i.e.,

$$\mathrm{wt}(\mathbf{x}) = d(\mathbf{x}, \mathbf{0}),$$

where $\mathbf{0}$ is the zero word.

Remark 4.3.2 For every element x of $\mathbf{F}_q$, we can define the Hamming weight as follows:

$$\mathrm{wt}(x) = d(x, 0) = \begin{cases} 1 & \text{if } x \neq 0 \\ 0 & \text{if } x = 0. \end{cases}$$

Table 4.1.

x	y	$x \star y$	$\mathrm{wt}(x) + \mathrm{wt}(y) - 2\mathrm{wt}(x \star y)$	$\mathrm{wt}(x + y)$
0	0	0	0	0
0	1	0	1	1
1	0	0	1	1
1	1	1	0	0

Then, writing $\mathbf{x} \in \mathbf{F}_q^n$ as $\mathbf{x} = (x_1, x_2, \ldots, x_n)$, the Hamming weight of $\mathbf{x}$ can also be equivalently defined as

$$\mathrm{wt}(\mathbf{x}) = \mathrm{wt}(x_1) + \mathrm{wt}(x_2) + \cdots + \mathrm{wt}(x_n). \tag{4.1}$$

Lemma 4.3.3 *If* $\mathbf{x}, \mathbf{y} \in \mathbf{F}_q^n$*, then* $d(\mathbf{x}, \mathbf{y}) = \mathrm{wt}(\mathbf{x} - \mathbf{y})$*.*

Proof. For $x, y \in \mathbf{F}_q$, $d(x, y) = 0$ if and only if $x = y$, which is true if and only if $x - y = 0$ or, equivalently, $\mathrm{wt}(x - y) = 0$. Lemma 4.3.3 now follows from (2.1) and (4.1). □

Since $a = -a$ for all $a \in \mathbf{F}_q$ when q is even, the following corollary is an immediate consequence of Lemma 4.3.3.

Corollary 4.3.4 *Let* q *be even. If* $\mathbf{x}, \mathbf{y} \in \mathbf{F}_q^n$*, then* $d(\mathbf{x}, \mathbf{y}) = \mathrm{wt}(\mathbf{x} + \mathbf{y})$*.*

For $\mathbf{x} = (x_1, x_2, \ldots, x_n)$ and $\mathbf{y} = (y_1, y_2, \ldots, y_n)$ in $\mathbf{F}_q^n$, let

$$\mathbf{x} \star \mathbf{y} = (x_1 y_1, x_2 y_2, \ldots, x_n y_n).$$

Lemma 4.3.5 *If* $\mathbf{x}, \mathbf{y} \in \mathbf{F}_2^n$*, then*

$$\mathrm{wt}(\mathbf{x} + \mathbf{y}) = \mathrm{wt}(\mathbf{x}) + \mathrm{wt}(\mathbf{y}) - 2\mathrm{wt}(\mathbf{x} \star \mathbf{y}). \tag{4.2}$$

Proof. From (4.1), it is enough to show that (4.2) is true for $\mathbf{x}, \mathbf{y} \in \mathbf{F}_2$. This can be easily verified as in Table 4.1. □

Clearly, Lemma 4.3.5 implies that $\mathrm{wt}(\mathbf{x}) + \mathrm{wt}(\mathbf{y}) \geq \mathrm{wt}(\mathbf{x} + \mathbf{y})$ for $\mathbf{x}, \mathbf{y} \in \mathbf{F}_2^n$. In fact, this inequality is true for any alphabet $\mathbf{F}_q$. The proof of the following lemma is left as an exercise (see Exercise 4.18).

Lemma 4.3.6 *For any prime power* q *and* $\mathbf{x}, \mathbf{y} \in \mathbf{F}_q^n$*, we have*

$$\mathrm{wt}(\mathbf{x}) + \mathrm{wt}(\mathbf{y}) \geq \mathrm{wt}(\mathbf{x} + \mathbf{y}) \geq \mathrm{wt}(\mathbf{x}) - \mathrm{wt}(\mathbf{y}). \tag{4.3}$$

Definition 4.3.7 Let C be a code (not necessarily linear). The *minimum (Hamming) weight* of C, denoted wt(C), is the smallest of the weights of the nonzero codewords of C.

Theorem 4.3.8 *Let C be a linear code over* $\mathbf{F}_q$. *Then* $d(C) = \mathrm{wt}(C)$.

Proof. Recall that for any words $\mathbf{x}$, $\mathbf{y}$ we have $d(\mathbf{x}, \mathbf{y}) = \mathrm{wt}(\mathbf{x} - \mathbf{y})$.

By definition, there exist $\mathbf{x}'$, $\mathbf{y}' \in C$ such that $d(\mathbf{x}', \mathbf{y}') = d(C)$, so

$$d(C) = d(\mathbf{x}', \mathbf{y}') = \mathrm{wt}(\mathbf{x}' - \mathbf{y}') \geq \mathrm{wt}(C),$$

since $\mathbf{x}' - \mathbf{y}' \in C$.

Conversely, there is a $\mathbf{z} \in C \setminus \{\mathbf{0}\}$ such that $\mathrm{wt}(C) = \mathrm{wt}(\mathbf{z})$, so

$$\mathrm{wt}(C) = \mathrm{wt}(\mathbf{z}) = d(\mathbf{z}, \mathbf{0}) \geq d(C). \qquad \square$$

Example 4.3.9 Consider the binary linear code $C = \{0000, 1000, 0100, 1100\}$. We see that

$$\mathrm{wt}(1000) = 1,$$
$$\mathrm{wt}(0100) = 1,$$
$$\mathrm{wt}(1100) = 2.$$

Hence, $d(C) = 1$.

Remark 4.3.10 (Some advantages of linear codes.) The following are some of the reasons why it may be preferable to use linear codes over nonlinear ones:

(i) As a linear code is a vector space, it can be described completely by using a basis (see Section 4.4).
(ii) The distance of a linear code is equal to the smallest weight of its nonzero codewords.
(iii) The encoding and decoding procedures for a linear code are faster and simpler than those for arbitrary nonlinear codes (see Sections 4.7 and 4.8).

4.4   Bases for linear codes

Since a linear code is a vector space, all its elements can be described in terms of a basis. In this section, we discuss three algorithms that yield either a basis for a given linear code or its dual. We first recall some facts from linear algebra.

Definition 4.4.1 Let A be a matrix over $\mathbf{F}_q$; an *elementary row operation* performed on A is any one of the following three operations:

(i) interchanging two rows,
(ii) multiplying a row by a nonzero scalar,
(iii) replacing a row by its sum with the scalar multiple of another row.

Definition 4.4.2 Two matrices are *row equivalent* if one can be obtained from the other by a sequence of elementary row operations.

The following are well known facts from linear algebra:

(i) Any matrix M over $\mathbf{F}_q$ can be put in *row echelon form (REF)* or *reduced row echelon form (RREF)* by a sequence of elementary row operations. In other words, a matrix is row equivalent to a matrix in REF or in RREF.
(ii) For a given matrix, its RREF is unique, but it may have different REFs. (Recall that the difference between the RREF and the REF is that the leading nonzero entry of a row in the RREF is equal to 1 and it is the only nonzero entry in its column.)

We are now ready to describe the three algorithms.

Algorithm 4.1

Input: A nonempty subset S of $\mathbf{F}_q^n$.

Output: A basis for $C = <S>$, the linear code generated by S.

Description: Form the matrix A whose rows are the words in S. Use elementary row operations to find an REF of A. Then the nonzero rows of the REF form a basis for C.

Example 4.4.3 Let $q = 3$. Find a basis for $C = <S>$, where

$$S = \{12101, 20110, 01122, 11010\}.$$

$$A = \begin{pmatrix} 12101 \\ 20110 \\ 01122 \\ 11010 \end{pmatrix} \rightarrow \begin{pmatrix} 12101 \\ 02211 \\ 01122 \\ 02212 \end{pmatrix} \rightarrow \begin{pmatrix} 12101 \\ 01122 \\ 00001 \\ 00000 \end{pmatrix}.$$

The last matrix is in REF. By Algorithm 4.1, $\{12101, 01122, 00001\}$ is a basis for C.

Algorithm 4.2

Input: A nonempty subset S of $\mathbf{F}_q^n$.

Output: A basis for $C = <S>$, the linear code generated by S.

Description: Form the matrix A whose columns are the words in S. Use elementary row operations to put A in REF and locate the leading columns in the REF. Then the original columns of A corresponding to these leading columns form a basis for C.

Example 4.4.4 Let $q = 2$. Find a basis for $C = <S>$, where

$$S = \{11101, 10110, 01011, 11010\}.$$

$$A = \begin{pmatrix} 1101 \\ 1011 \\ 1100 \\ 0111 \\ 1010 \end{pmatrix} \rightarrow \begin{pmatrix} 1101 \\ 0110 \\ 0001 \\ 0111 \\ 0111 \end{pmatrix} \rightarrow \begin{pmatrix} 1101 \\ 0110 \\ 0001 \\ 0000 \\ 0000 \end{pmatrix}.$$

Since columns 1, 2 and 4 of the REF are the leading columns, Algorithm 4.2 says that columns 1, 2 and 4 of A form a basis for C; i.e., $\{11101, 10110, 11010\}$ is a basis for C.

Remark 4.4.5 Note that the basis that Algorithm 4.2 yields is a subset of the given set S, while this is not necessarily the case for Algorithm 4.1.

Algorithm 4.3

Input: A nonempty subset S of $\mathbf{F}_q^n$.

Output: A basis for the dual code $C^\perp$, where $C = <S>$.

Description: Form the matrix A whose rows are the words in S. Use elementary row operations to place A in RREF. Let G be the $k \times n$ matrix consisting of all the nonzero rows of the RREF:

$$A \rightarrow \begin{pmatrix} G \\ O \end{pmatrix}.$$

(Here, O denotes the zero matrix.)

The matrix G contains k leading columns. Permute the columns of G to form

$$G' = (I_k | X),$$

where I_k denotes the $k \times k$ identity matrix. Form a matrix H' as follows:

$$H' = \left(-X^\mathrm{T}|I_{n-k}\right),$$

where X^T denotes the transpose of X.

Apply the inverse of the permutation applied to the columns of G to the columns of H' to form H. Then the rows of H form a basis for $C^\perp$.

Remark 4.4.6 (i) Notice that Algorithm 4.3 also provides a basis for C since it includes Algorithm 4.1.

(ii) An explanation of the principles behind Algorithm 4.3 is given in Theorem 4.5.9 in the following section.

Example 4.4.7 Let $q = 3$. Find a basis for $C^\perp$ if the RREF of A is

$$G = \begin{array}{c} \begin{array}{cccccccccc} 1 & 2 & 3 & 4 & 5 & 6 & 7 & 8 & 9 & 10 \end{array} \\ \left(\begin{array}{cccccccccc} 1 & 0 & 2 & 0 & 0 & 2 & 0 & 1 & 0 & 2 \\ 0 & 0 & 0 & 1 & 0 & 1 & 0 & 0 & 0 & 1 \\ 0 & 0 & 0 & 0 & 1 & 0 & 0 & 2 & 0 & 0 \\ 0 & 0 & 0 & 0 & 0 & 0 & 1 & 0 & 0 & 1 \\ 0 & 0 & 0 & 0 & 0 & 0 & 0 & 0 & 1 & 2 \end{array}\right). \end{array}$$

The leading columns of G are columns $1, 4, 5, 7$ and 9. We permute the columns of G into the order $1, 4, 5, 7, 9, 2, 3, 6, 8, 10$ to form the matrix

$$G' = (I_5|X) = \begin{array}{c} \begin{array}{cccccccccc} 1 & 4 & 5 & 7 & 9 & 2 & 3 & 6 & 8 & 10 \end{array} \\ \left(\begin{array}{cccccccccc} 1 & 0 & 0 & 0 & 0 & 0 & 2 & 2 & 1 & 2 \\ 0 & 1 & 0 & 0 & 0 & 0 & 0 & 1 & 0 & 1 \\ 0 & 0 & 1 & 0 & 0 & 0 & 0 & 0 & 2 & 0 \\ 0 & 0 & 0 & 1 & 0 & 0 & 0 & 0 & 0 & 1 \\ 0 & 0 & 0 & 0 & 1 & 0 & 0 & 0 & 0 & 2 \end{array}\right). \end{array}$$

Form the matrix H' and finally rearrange the columns of H' using the inverse permutation to obtain H:

$$H' = \begin{array}{c} \begin{array}{cccccccccc} 1 & 4 & 5 & 7 & 9 & 2 & 3 & 6 & 8 & 10 \end{array} \\ \left(\begin{array}{cccccccccc} 0 & 0 & 0 & 0 & 0 & 1 & 0 & 0 & 0 & 0 \\ 1 & 0 & 0 & 0 & 0 & 0 & 1 & 0 & 0 & 0 \\ 1 & 2 & 0 & 0 & 0 & 0 & 0 & 1 & 0 & 0 \\ 2 & 0 & 1 & 0 & 0 & 0 & 0 & 0 & 1 & 0 \\ 1 & 2 & 0 & 2 & 1 & 0 & 0 & 0 & 0 & 1 \end{array}\right), \end{array}$$

$$
\begin{array}{cccccccccc}
1 & 2 & 3 & 4 & 5 & 6 & 7 & 8 & 9 & 10
\end{array}
$$

$$
H = \begin{pmatrix}
0 & 1 & 0 & 0 & 0 & 0 & 0 & 0 & 0 & 0 \\
1 & 0 & 1 & 0 & 0 & 0 & 0 & 0 & 0 & 0 \\
1 & 0 & 0 & 2 & 0 & 1 & 0 & 0 & 0 & 0 \\
2 & 0 & 0 & 0 & 1 & 0 & 0 & 1 & 0 & 0 \\
1 & 0 & 0 & 2 & 0 & 0 & 2 & 0 & 1 & 1
\end{pmatrix}.
$$

By Algorithm 4.3, the rows of H form a basis for $C^{\perp}$.

4.5 Generator matrix and parity-check matrix

Knowing a basis for a linear code enables us to describe its codewords explicitly. In coding theory, a basis for a linear code is often represented in the form of a matrix, called a generator matrix, while a matrix that represents a basis for the dual code is called a parity-check matrix. These matrices play an important role in coding theory.

Definition 4.5.1 (i) A *generator matrix* for a linear code C is a matrix G whose rows form a basis for C.

(ii) A *parity-check matrix* H for a linear code C is a generator matrix for the dual code $C^{\perp}$.

Remark 4.5.2 (i) If C is an $[n, k]$-linear code, then a generator matrix for C must be a $k \times n$ matrix and a parity-check matrix for C must be an $(n - k) \times n$ matrix.

(ii) Algorithm 4.3 of Section 4.4 can be used to find generator and parity-check matrices for a linear code.

(iii) As the number of bases for a vector space usually exceeds one, the number of generator matrices for a linear code also usually exceeds one. Moreover, even when the basis is fixed, a permutation (different from the identity) of the rows of a generator matrix also leads to a different generator matrix.

(iv) The rows of a generator matrix are linearly independent. The same holds for the rows of a parity-check matrix. To show that a $k \times n$ matrix G is indeed a generator matrix for a given $[n, k]$-linear code C, it suffices to show that the rows of G are codewords in C and that they are linearly independent. Alternatively, one may also show that C is contained in the row space of G.

Definition 4.5.3 (i) A generator matrix of the form $(I_k | X)$ is said to be in *standard form*.

(ii) A parity-check matrix in the form $(Y | I_{n-k})$ is said to be in *standard form*.

Lemma 4.5.4 *Let C be an $[n, k]$-linear code over $\mathbf{F}_q$, with generator matrix G. Then $\mathbf{v} \in \mathbf{F}_q^n$ belongs to $C^\perp$ if and only if $\mathbf{v}$ is orthogonal to every row of G; i.e., $\mathbf{v} \in C^\perp \Leftrightarrow \mathbf{v}G^\mathrm{T} = \mathbf{0}$. In particular, given an $(n - k) \times n$ matrix H, then H is a parity-check matrix for C if and only if the rows of H are linearly independent and $HG^\mathrm{T} = O$.*

Proof. Let $\mathbf{r}_i$ denote the ith row of G. In particular, $\mathbf{r}_i \in C$ for all $1 \le i \le k$, and every $\mathbf{c} \in C$ may be written as

$$\mathbf{c} = \lambda_1 \mathbf{r}_1 + \cdots + \lambda_k \mathbf{r}_k,$$

where $\lambda_1, \ldots, \lambda_k \in \mathbf{F}_q$.

If $\mathbf{v} \in C^\perp$, then $\mathbf{v} \cdot \mathbf{c} = 0$ for all $\mathbf{c} \in C$. In particular, $\mathbf{v}$ is orthogonal to $\mathbf{r}_i$, for all $1 \le i \le k$; i.e., $\mathbf{v}G^\mathrm{T} = \mathbf{0}$.

Conversely, if $\mathbf{v} \cdot \mathbf{r}_i = 0$ for all $1 \le i \le k$, then clearly, for any $\mathbf{c} = \lambda_1 \mathbf{r}_1 + \cdots + \lambda_k \mathbf{r}_k \in C$,

$$\mathbf{v} \cdot \mathbf{c} = \lambda_1 (\mathbf{v} \cdot \mathbf{r}_1) + \cdots + \lambda_k (\mathbf{v} \cdot \mathbf{r}_k) = 0.$$

For the last statement, if H is a parity-check matrix for C, then the rows of H are linearly independent by definition. Since the rows of H are codewords in $C^\perp$, it follows from the earlier statement that $HG^\mathrm{T} = O$.

Conversely, if $HG^\mathrm{T} = O$, then the earlier statement shows that the rows of H, and hence the row space of H, are contained in $C^\perp$. Since the rows of H are linearly independent, the row space of H has dimension $n - k$, so the row space of H is indeed $C^\perp$. In other words, H is a parity-check matrix for C. $\square$

Remark 4.5.5 An alternative but equivalent formulation for Lemma 4.5.4 is the following:

Let C be an $[n, k]$-linear code over $\mathbf{F}_q$, with parity-check matrix H. Then $\mathbf{v} \in \mathbf{F}_q^n$ belongs to C if and only if $\mathbf{v}$ is orthogonal to every row of H; i.e., $\mathbf{v} \in C \Leftrightarrow \mathbf{v}H^\mathrm{T} = \mathbf{0}$. In particular, given a $k \times n$ matrix G, then G is a generator matrix for C if and only if the rows of G are linearly independent and $GH^\mathrm{T} = O$.

One of the consequences of Lemma 4.5.4 is the following theorem relating the distance d of a linear code C to properties of a parity-check matrix of C. When d is small, Corollary 4.5.7 can be a useful way to determine d.

Theorem 4.5.6 *Let C be a linear code and let H be a parity-check matrix for C. Then*

(i) C has distance $\geq d$ if and only if any $d - 1$ columns of H are linearly independent; and

(ii) C has distance $\leq d$ if and only if H has d columns that are linearly dependent.

Proof. Let $\mathbf{v} = (v_1, \ldots, v_n) \in C$ be a word of weight $e > 0$. Suppose the nonzero coordinates are in the positions $i_1, \ldots, i_e$, so that $v_j = 0$ if $j \notin \{i_1, \ldots, i_e\}$. Let $\mathbf{c}_i$ $(1 \leq i \leq n)$ denote the ith column of H.

By Lemma 4.5.4 (or, more precisely, its equivalent formulation in Remark 4.5.5), C contains a nonzero word $\mathbf{v} = (v_1, \ldots, v_n)$ of weight e (whose nonzero coordinates are $v_{i_1}, \ldots, v_{i_e}$) if and only if

$$\mathbf{0} = \mathbf{v}H^{\mathrm{T}} = v_{i_1}\mathbf{c}_{i_1}^{\mathrm{T}} + \cdots + v_{i_e}\mathbf{c}_{i_e}^{\mathrm{T}},$$

which is true if and only if there are e columns of H (namely, $\mathbf{c}_{i_1}, \ldots, \mathbf{c}_{i_e}$) that are linearly dependent.

To say that the distance of C is $\geq d$ is equivalent to saying that C does not contain any nonzero word of weight $\leq d - 1$, which is in turn equivalent to saying that any $\leq d-1$ columns of H are linearly independent. This proves (i).

Similarly, to say that the distance of C is $\leq d$ is equivalent to saying that C contains a nonzero word of weight $\leq d$, which is in turn equivalent to saying that H has $\leq d$ columns (and hence d columns) that are linearly dependent. This proves (ii). $\qquad\square$

An immediate corollary of Theorem 4.5.6 is the following result.

Corollary 4.5.7 *Let C be a linear code and let H be a parity-check matrix for C. Then the following statements are equivalent:*

(i) *C has distance d;*

(ii) *any $d - 1$ columns of H are linearly independent and H has d columns that are linearly dependent.*

Example 4.5.8 Let C be the binary linear code with parity-check matrix

$$H = \begin{pmatrix} 10100 \\ 11010 \\ 01001 \end{pmatrix}.$$

By inspection, it is seen that there are no zero columns and no two columns of H sum to $\mathbf{0}^{\mathrm{T}}$, so any two columns of H are linearly independent. However, columns 1, 3 and 4 sum to $\mathbf{0}^{\mathrm{T}}$, and hence are linearly dependent. Therefore, the distance of C is $d = 3$.

Theorem 4.5.9 *If* $G = (I_k|X)$ *is the standard form generator matrix of an* $[n, k]$-*code* C, *then a parity-check matrix for* C *is* $H = (-X^T|I_{n-k})$.

Proof. Obviously, the equation $HG^T = O$ is satisfied. By considering the last $n - k$ coordinates, it is clear that the rows of H are linearly independent. Therefore, the conclusion follows from Lemma 4.5.4. □

Remark 4.5.10 Theorem 4.5.9 shows that Algorithm 4.3 of Section 4.4 actually gives what it claims to yield.

Example 4.5.11 Find a generator matrix and a parity-check matrix for the binary linear code $C = <S>$, where $S = \{11101, 10110, 01011, 11010\}$.
By Algorithm 4.1,

$$A = \begin{pmatrix} 11101 \\ 10110 \\ 01011 \\ 11010 \end{pmatrix} \rightarrow \begin{pmatrix} 11101 \\ 01011 \\ 00111 \\ 00000 \end{pmatrix} \rightarrow \begin{pmatrix} 10001 \\ 01011 \\ 00111 \\ 00000 \end{pmatrix},$$

which is in RREF. By Algorithm 4.3, we have

$$G = \begin{pmatrix} 100|01 \\ 010|11 \\ 001|11 \end{pmatrix}, \qquad H = \begin{pmatrix} 0 & 1 & 1 & 1 & 0 \\ 1 & 1 & 1 & 0 & 1 \end{pmatrix}.$$

Here, G is a generator matrix for C and H is a parity-check matrix for C. We can verify that $GH^T = O = HG^T$.

It should be noted that it is not true that every linear code has a generator matrix in standard form.

Example 4.5.12 Consider the binary linear code

$$C = \{000, 001, 100, 101\}.$$

Since $\dim(C) = 2$, by Theorem 4.1.15(ii) the number of bases for C is

$$\frac{1}{2!}(2^2 - 1)(2^2 - 2) = 3.$$

We can list all the bases for C:

$$\{001, 100\}, \quad \{001, 101\}, \quad \{100, 101\}.$$

Hence, C has six generator matrices:

$$\begin{pmatrix} 001 \\ 100 \end{pmatrix}, \begin{pmatrix} 100 \\ 001 \end{pmatrix}, \begin{pmatrix} 001 \\ 101 \end{pmatrix}, \begin{pmatrix} 101 \\ 001 \end{pmatrix}, \begin{pmatrix} 100 \\ 101 \end{pmatrix}, \begin{pmatrix} 101 \\ 100 \end{pmatrix}.$$

Note that none of these matrices is in standard form.

4.6   Equivalence of linear codes

While certain linear codes may not have a generator matrix in standard form, after a suitable permutation of the coordinates of the codewords and possibly multiplying certain coordinates with some nonzero scalars, one can always arrive at a new code which has a generator matrix in standard form.

Definition 4.6.1 Two (n, M)-codes over $\mathbf{F}_q$ are *equivalent* if one can be obtained from the other by a combination of operations of the following types:

(i) permutation of the n digits of the codewords;
(ii) multiplication of the symbols appearing in a fixed position by a nonzero scalar.

Example 4.6.2 (i) Let $q = 2$ and $n = 4$. Choosing to rearrange the bits in the order 2, 4, 1, 3, we see that the code

$$C = \{0000, 0101, 0010, 0111\}$$

is equivalent to the code

$$C' = \{0000, 1100, 0001, 1101\}.$$

(ii) Let $q = 3$ and $n = 3$. Consider the ternary code

$$C = \{000, 011, 022\}.$$

Permuting the first and second positions, followed by multiplying the third position by 2, we obtain the equivalent code

$$C' = \{000, 102, 201\}.$$

Theorem 4.6.3 *Any linear code C is equivalent to a linear code C' with a generator matrix in standard form.*

Proof. If G is a generator matrix for C, place G in RREF. Rearrange the columns of the RREF so that the leading columns come first and form an identity matrix.

The result is a matrix, G', in standard form which is a generator matrix for a code C' equivalent to the code C. □

Remark 4.6.4 Theorem 4.6.3 is essentially the first part of Algorithm 4.3 of Section 4.4.

Example 4.6.5 Let C be a binary linear code with generator matrix

$$G = \begin{pmatrix} 1100001 \\ 0010011 \\ 0001001 \end{pmatrix}.$$

Rearranging the columns in the order 1, 3, 4, 2, 5, 6, 7 yields the matrix

$$G' = \begin{pmatrix} 100 \mid 1001 \\ 010 \mid 0011 \\ 001 \mid 0001 \end{pmatrix}.$$

Let C' be the code generated by G'; then C' is equivalent to C and C' has a generator matrix G', which is in standard form.

Example 4.6.6 We saw in Example 4.5.12 that the binary linear code $C = \{000, 001, 100, 101\}$ does not have a generator matrix in standard form. However, if we permute the second and third coordinates, we obtain the equivalent binary linear code

$$C' = \{000, 010, 100, 110\},$$

and it is clear that

$$\begin{pmatrix} 100 \\ 010 \end{pmatrix}$$

is a generator matrix in standard form for C'.

4.7 Encoding with a linear code

Let C be an $[n, k, d]$-linear code over the finite field $\mathbf{F}_q$. Each codeword of C can represent one piece of information, so C can represent q^k distinct pieces of information. Once a basis $\{\mathbf{r}_1, \ldots, \mathbf{r}_k\}$ is fixed for C, each codeword $\mathbf{v}$, or, equivalently, each of the q^k pieces of information, can be uniquely written as a linear combination,

$$\mathbf{v} = u_1\mathbf{r}_1 + \cdots + u_k\mathbf{r}_k,$$

where $u_1, \ldots, u_k \in \mathbf{F}_q$.

Equivalently, we may set G to be the generator matrix of C whose ith row is the vector $\mathbf{r}_i$ in the chosen basis. Given a vector $\mathbf{u} = (u_1, \ldots, u_k) \in \mathbf{F}_q^k$, it is clear that

$$\mathbf{v} = \mathbf{u}G = u_1\mathbf{r}_1 + \cdots + u_k\mathbf{r}_k$$

is a codeword in C. Conversely, any $\mathbf{v} \in C$ can be written uniquely as $\mathbf{v} = \mathbf{u}G$, where $\mathbf{u} = (u_1, \ldots, u_k) \in \mathbf{F}_q^k$. Hence, every word $\mathbf{u} \in \mathbf{F}_q^k$ can be encoded as $\mathbf{v} = \mathbf{u}G$.

The process of representing the elements $\mathbf{u}$ of $\mathbf{F}_q^k$ as codewords $\mathbf{v} = \mathbf{u}G$ in C is called *encoding*.

Example 4.7.1 Let C be the binary [5, 3]-linear code with the generator matrix

$$G = \begin{pmatrix} 10110 \\ 01011 \\ 00101 \end{pmatrix};$$

then the message $\mathbf{u} = 101$ is encoded as

$$\mathbf{v} = \mathbf{u}G = (101)\begin{pmatrix} 10110 \\ 01011 \\ 00101 \end{pmatrix} = 10011.$$

Note that the information rate of C is $3/5$, i.e., only 3 bits out of 5 are used to carry the message.

Remark 4.7.2 (Advantages of having G in standard form.)   Some of the advantages of having the generator matrix of a linear code in standard form are as follows:

(i) If a linear code C has a generator matrix G in standard form, $G = (I|X)$, then Algorithm 4.3 of Section 4.4 at once yields

$$H = (-X^T|I)$$

as a parity-check matrix for C.

(ii) If an $[n, k, d]$-linear code C has a generator matrix G in standard form, $G = (I|X)$, then it is trivial to recover the message $\mathbf{u}$ from the codeword $\mathbf{v} = \mathbf{u}G$ since

$$\mathbf{v} = \mathbf{u}G = \mathbf{u}(I|X) = (\mathbf{u}, \mathbf{u}X);$$

i.e., the first k digits in the codeword $\mathbf{v} = \mathbf{u}G$ give the message $\mathbf{u}$ – they are called the *message digits*. The remaining $n - k$ digits are called *check digits.* The check digits represent the *redundancy* which has been added to the message for protection against noise.

4.8   Decoding of linear codes

A code is of practical use only if an efficient decoding scheme can be applied to it. In this section, we discuss a rather simple but elegant nearest neighbour decoding for linear codes, as well as a modification that improves its performance when the length of the code is large.

4.8.1   Cosets

We begin with the notion of a coset. Cosets play a crucial role in the decoding schemes to be discussed in this chapter.

Definition 4.8.1 Let C be a linear code of length n over $\mathbf{F}_q$, and let $\mathbf{u} \in \mathbf{F}_q^n$ be any vector of length n; we define the *coset* of C determined by $\mathbf{u}$ to be the set

$$C + \mathbf{u} = \{\mathbf{v} + \mathbf{u} : \mathbf{v} \in C\} (= \mathbf{u} + C).$$

Remark 4.8.2 For the reader who knows some group theory, note that, by considering the vector addition, $\mathbf{F}_q^n$ is a finite abelian group, and a linear code C over $\mathbf{F}_q$ of length n is also a subgroup of $\mathbf{F}_q^n$. The coset of a linear code defined above coincides with the usual notion of a coset in group theory.

Example 4.8.3 Let $q = 2$ and $C = \{000, 101, 010, 111\}$. Then

$$C + 000 = \{000, 101, 010, 111\},$$
$$C + 001 = \{001, 100, 011, 110\},$$
$$C + 010 = \{010, 111, 000, 101\},$$
$$C + 011 = \{011, 110, 001, 100\},$$
$$C + 100 = \{100, 001, 110, 011\},$$
$$C + 101 = \{101, 000, 111, 010\},$$
$$C + 110 = \{110, 011, 100, 001\},$$
$$C + 111 = \{111, 010, 101, 000\}.$$

Note that

$$C + 000 = C + 010 = C + 101 = C + 111 = C;$$
$$C + 001 = C + 011 = C + 100 = C + 110 = \mathbf{F}_2^3 \backslash C.$$

Theorem 4.8.4 *Let C be an $[n, k, d]$-linear code over the finite field $\mathbf{F}_q$. Then,*

(i) *every vector of $\mathbf{F}_q^n$ is contained in some coset of C;*

(ii) *for all $\mathbf{u} \in \mathbf{F}_q^n$, $|C + \mathbf{u}| = |C| = q^k$;*

(iii) *for all $\mathbf{u}, \mathbf{v} \in \mathbf{F}_q^n$, $\mathbf{u} \in C + \mathbf{v}$ implies that $C + \mathbf{u} = C + \mathbf{v}$;*

(iv) *two cosets are either identical or they have empty intersection;*

(v) *there are q^{n-k} different cosets of C;*

(vi) *for all $\mathbf{u}, \mathbf{v} \in \mathbf{F}_q^n$, $\mathbf{u} - \mathbf{v} \in C$ if and only if $\mathbf{u}$ and $\mathbf{v}$ are in the same coset.*

Proof. (i) The vector $\mathbf{v} \in \mathbf{F}_q^n$ is clearly contained in the coset $C + \mathbf{v}$.

(ii) By definition, $C + \mathbf{u}$ has at most $|C| = q^k$ elements. Clearly, two elements $\mathbf{c} + \mathbf{u}$ and $\mathbf{c}' + \mathbf{u}$ of $C + \mathbf{u}$ are equal if and only if $\mathbf{c} = \mathbf{c}'$, hence $|C + \mathbf{u}| = |C| = q^k$.

(iii) It follows from the definition of $C + \mathbf{v}$ that $C + \mathbf{u} \subseteq C + \mathbf{v}$. Then, by (ii), $C + \mathbf{u} = C + \mathbf{v}$.

(iv) Consider two cosets $C + \mathbf{u}$ and $C + \mathbf{v}$ and suppose $\mathbf{x} \in (C + \mathbf{u}) \cap (C + \mathbf{v})$. Since $\mathbf{x} \in C + \mathbf{u}$, (iii) shows that $C + \mathbf{u} = C + \mathbf{x}$. Similarly, since $\mathbf{x} \in C + \mathbf{v}$, it follows that $C + \mathbf{v} = C + \mathbf{x}$. Hence, $C + \mathbf{u} = C + \mathbf{v}$.

(v) follows immediately from (i), (ii) and (iv).

(vi) If $\mathbf{u} - \mathbf{v} = \mathbf{c} \in C$, then $\mathbf{u} = \mathbf{c} + \mathbf{v} \in C + \mathbf{v}$, so $C + \mathbf{u} = C + \mathbf{v}$. By the proof of (i), $\mathbf{u} \in C + \mathbf{u}$ and $\mathbf{v} \in C + \mathbf{v}$, so $\mathbf{u}$ and $\mathbf{v}$ are in the same coset.

Conversely, suppose $\mathbf{u}, \mathbf{v}$ are both in the coset $C + \mathbf{x}$. Then $\mathbf{u} = \mathbf{c} + \mathbf{x}$ and $\mathbf{v} = \mathbf{c}' + \mathbf{x}$, for some $\mathbf{c}, \mathbf{c}' \in C$. Hence, $\mathbf{u} - \mathbf{v} = \mathbf{c} - \mathbf{c}' \in C$. $\square$

Example 4.8.5 The cosets of the binary linear code

$$C = \{0000, 1011, 0101, 1110\}$$

are as follows:

$0000 + C$:	0000	1011	0101	1110
$0001 + C$:	0001	1010	0100	1111
$0010 + C$:	0010	1001	0111	1100
$1000 + C$:	1000	0011	1101	0110

Remark 4.8.6 The above array is called a (*Slepian*) *standard array*.

Definition 4.8.7 A word of the least (Hamming) weight in a coset is called a *coset leader*.

Example 4.8.8 In Example 4.8.5, the vector $\mathbf{u}$ in $\mathbf{u} + C$ of the first column are coset leaders for the respective cosets. Note that the coset $0001 + C$ can also have as coset leader 0100.

4.8.2   Nearest neighbour decoding for linear codes

Let C be a linear code. Assume the codeword $\mathbf{v}$ is transmitted and the word $\mathbf{w}$ is received, resulting in the *error pattern* (or *error string*)

$$\mathbf{e} = \mathbf{w} - \mathbf{v} \in \mathbf{w} + C.$$

Then $\mathbf{w} - \mathbf{e} = \mathbf{v} \in C$, so, by part (vi) of Theorem 4.8.4, the error pattern $\mathbf{e}$ and the received word $\mathbf{w}$ are in the same coset.

Since error patterns of small weight are the most likely to occur, nearest neighbour decoding works for a linear code C in the following manner. Upon receiving the word $\mathbf{w}$, we choose a word $\mathbf{e}$ of least weight in the coset $\mathbf{w} + C$ and conclude that $\mathbf{v} = \mathbf{w} - \mathbf{e}$ was the codeword transmitted.

Example 4.8.9 Let $q = 2$ and $C = \{0000, 1011, 0101, 1110\}$. Decode the following received words: (i) $\mathbf{w} = 1101$; (ii) $\mathbf{w} = 1111$.

First, we write down the standard array of C (exactly the one in Example 4.8.5):

$0000 + C$:	0000	1011	0101	1110
$0001 + C$:	0001	1010	0100	1111
$0010 + C$:	0010	1001	0111	1100
$1000 + C$:	1000	0011	1101	0110

(i) $\mathbf{w} = 1101$: $\mathbf{w} + C$ is the fourth coset. The word of least weight in this coset is 1000 (note that this is the unique coset leader of this coset). Hence, $1101 - 1000 = 1101 + 1000 = 0101$ was the most likely codeword transmitted (note that this is the word at the top of the column where the received word 1101 is found).

(ii) $\mathbf{w} = 1111$: $\mathbf{w} + C$ is the second coset. There are two words of smallest weight, 0001 and 0100, in this coset. (This means that there are two choices for the coset leader. In the array above, we have chosen 0001 as the coset leader. If we had chosen 0100, we would have obtained a slightly different array.) When the coset of the received word has more than one possible leader, the approach we take for decoding depends on the decoding scheme (i.e., in-complete or complete) used. If we are doing incomplete decoding, we ask for a retransmission. If we are doing complete decoding, we arbitrarily choose one of the words of smallest weight, say 0001, to be the error pattern, and conclude that $1111 - 0001 = 1111 + 0001 = 1110$ was a most likely codeword sent. (Note: this means we choose 0001 as the coset leader, form the standard array as above, then observe that a most likely word sent is again found at the top of the column where the received word is located.) What happens if we choose 0100 as the coset leader/error pattern?

4.8.3 Syndrome decoding

The decoding scheme based on the standard array works reasonably well when the length n of the linear code is small, but it may take a considerable amount of time when n is large. Some time can be saved by making use of the syndrome to identify the coset to which the received word belongs.

Definition 4.8.10 Let C be an $[n, k, d]$-linear code over $\mathbf{F}_q$ and let H be a parity-check matrix for C. For any $\mathbf{w} \in \mathbf{F}_q^n$, the *syndrome* of $\mathbf{w}$ is the word $S(\mathbf{w}) = \mathbf{w}H^{\mathrm{T}} \in \mathbf{F}_q^{n-k}$. (Strictly speaking, as the syndrome depends on the choice of the parity-check matrix H, it is more appropriate to denote the syndrome of $\mathbf{w}$ by $S_H(\mathbf{w})$ to emphasize this dependence. However, for simplicity of notation, the suffix H is dropped whenever there is no risk of ambiguity.)

Theorem 4.8.11 *Let C be an $[n, k, d]$-linear code and let H be a parity-check matrix for C. For $\mathbf{u}, \mathbf{v} \in \mathbf{F}_q^n$, we have*

(i) $S(\mathbf{u} + \mathbf{v}) = S(\mathbf{u}) + S(\mathbf{v})$;
(ii) $S(\mathbf{u}) = \mathbf{0}$ *if and only if $\mathbf{u}$ is a codeword in C*;
(iii) $S(\mathbf{u}) = S(\mathbf{v})$ *if and only if $\mathbf{u}$ and $\mathbf{v}$ are in the same coset of C.*

Proof. (i) is an immediate consequence of the definition of the syndrome.

(ii) By the definition of the syndrome, $S(\mathbf{u}) = \mathbf{0}$ if and only if $\mathbf{u}H^{\mathrm{T}} = \mathbf{0}$, which, by Remark 4.5.5, is equivalent to $\mathbf{u} \in C$.

(iii) follows from (i), (ii) and Theorem 4.8.4(vi).   □

Remark 4.8.12 (i) Part (iii) of Theorem 4.8.11 says that we can identify a coset by its syndrome; conversely, all the words in a given coset yield the same syndrome, so the syndrome of a coset is the syndrome of any word in the coset. In other words, there is a one-to-one correspondence between the cosets and the syndromes.

(ii) Since the syndromes are in $\mathbf{F}_q^{n-k}$, there are at most q^{n-k} syndromes. Theorem 4.8.4(v) says that there are q^{n-k} cosets, so there are q^{n-k} corresponding syndromes (all distinct). Therefore, *all* the vectors in $\mathbf{F}_q^{n-k}$ appear as syndromes.

Definition 4.8.13 A table which matches each coset leader with its syndrome is called a *syndrome look-up table*. (Sometimes such a table is called a *standard decoding array (SDA)*.)

Table 4.2.

Coset leader $\mathbf{u}$	Syndrome $S(\mathbf{u})$
0000	00
0001	01
0010	10
1000	11

Steps to construct a syndrome look-up table assuming complete nearest neighbour decoding

Step 1: List all the cosets for the code, choose from each coset a word of least weight as coset leader $\mathbf{u}$.

Step 2: Find a parity-check matrix H for the code and, for each coset leader $\mathbf{u}$, calculate its syndrome $S(\mathbf{u}) = \mathbf{u}H^{\mathrm{T}}$.

Remark 4.8.14 For incomplete nearest neighbour decoding, if we find more than one word of smallest weight in Step 1 of the above procedure, place the symbol '$*$' in that entry of the syndrome look-up table to indicate that retransmission is required.

Example 4.8.15 Assume complete nearest neighbour decoding. Construct a syndrome look-up table for the binary linear code

$$C = \{0000, 1011, 0101, 1110\}.$$

From the cosets computed earlier, we choose the words 0000, 0001, 0010 and 1000 as coset leaders. Next, a parity-check matrix for C is

$$H = \begin{pmatrix} 1 & 0 & 1 & 0 \\ 1 & 1 & 0 & 1 \end{pmatrix}.$$

Now we construct a syndrome look-up table for C (Table 4.2). (We may also interchange the two columns.) Note that each word of length 2 occurs exactly once as a syndrome.

Example 4.8.16 A syndrome look-up table for C, assuming incomplete nearest neighbour decoding, is given in Table 4.3.

Remark 4.8.17 (i) Note that a unique coset leader corresponds to an error pattern that can be corrected, assuming incomplete nearest neighbour decoding.

Table 4.3.

Coset leader **u**	Syndrome $S(\mathbf{u})$
0000	00
*	01
0010	10
1000	11

A coset leader (not necessarily unique) corresponds to an error pattern that can be corrected, assuming complete nearest neighbour decoding.

(ii) A quicker way to construct a syndrome look-up table, given the parity-check matrix H and distance d for the code C, is to generate all the error patterns **e** with

$$\text{wt}(\mathbf{e}) \leq \left\lfloor \frac{d-1}{2} \right\rfloor$$

as coset leaders (cf. Exercise 4.44) and compute the syndrome $S(\mathbf{e})$ for each of them.

Example 4.8.18 Assuming complete nearest neighbour decoding, construct a syndrome look-up table for the binary linear code C with parity-check matrix H, where

$$H = \begin{pmatrix} 1 & 0 & 1 & 1 & 0 & 0 \\ 1 & 1 & 1 & 0 & 1 & 0 \\ 0 & 1 & 1 & 0 & 0 & 1 \end{pmatrix}.$$

First, we claim that the distance of C is $d = 3$. This can be easily seen by applying Corollary 4.5.7 and observing that no two columns of H are linearly dependent while the second, third and fourth columns are linearly dependent.

As $\lfloor (d-1)/2 \rfloor = 1$, all the error patterns with weight 0 or 1 will be coset leaders. We then compute the syndrome for each of them and obtain the first seven rows of the syndrome look-up table. Since every word of length 3 must occur as a syndrome, the remaining coset leader **u** has syndrome $\mathbf{u}H^{\text{T}} = 101$. Moreover, **u** must have weight ≥ 2 since all the words of weight 0 or 1 have already been included in the syndrome look-up table. Since we are looking for a coset leader, it is reasonable to start looking among the remaining words of the smallest available weight, i.e., 2. Doing so, we find three possible coset leaders: 000101, 001010 and 110000. Since we are using complete nearest neighbour decoding, we can arbitrarily choose 000101 as a coset leader and complete the syndrome look-up table (Table 4.4).

Table 4.4.

Coset leader $\mathbf{u}$	Syndrome $S(\mathbf{u})$
000000	000
100000	110
010000	011
001000	111
000100	100
000010	010
000001	001
000101	101

Table 4.2. Repeated   from
p. 63.

Coset leader $\mathbf{u}$	Syndrome $S(\mathbf{u})$
0000	00
0001	01
0010	10
1000	11

Note that, if incomplete nearest neighbour decoding is used, the coset leader 000101 in the last row of Table 4.4 will be replaced by '$*$'.

Decoding procedure for syndrome decoding

Step 1: For the received word $\mathbf{w}$, compute the syndrome $S(\mathbf{w})$.
Step 2: Find the coset leader $\mathbf{u}$ next to the syndrome $S(\mathbf{w}) = S(\mathbf{u})$ in the syndrome look-up table.
Step 3: Decode $\mathbf{w}$ as $\mathbf{v} = \mathbf{w} - \mathbf{u}$.

Example 4.8.19 Let $q = 2$ and let $C = \{0000, 1011, 0101, 1110\}$. Use the syndrome look-up table constructed in Example 4.8.15 to decode (i) $\mathbf{w} = 1101$; (ii) $\mathbf{w} = 1111$.

Recall the syndrome look-up table constructed in Example 4.8.15 (Table 4.2, repeated here for convenience).

(i) $\mathbf{w} = 1101$. The syndrome is $S(\mathbf{w}) = \mathbf{w}H^{\mathrm{T}} = 11$. From Table 4.2, we see that the coset leader is 1000. Hence, $1101 + 1000 = 0101$ was a most likely codeword sent.

(ii) $\mathbf{w} = 1111$. The syndrome is $S(\mathbf{w}) = \mathbf{w}H^{\mathrm{T}} = 01$. From Table 4.2, we see that the coset leader is 0001. Hence, $1111 + 0001 = 1110$ was a most likely codeword sent.

Exercises

4.1 Prove Proposition 4.1.6.

4.2 For each of the following sets, determine whether it is a vector space over the given finite field $\mathbf{F}_q$. If it is a vector space, determine the number of distinct bases it can have.

(a) $q = 2$, $S = \{(a, b, c, d, e) \ : \ a + b + c + d + e = 1\}$,

(b) $q = 3$, $T = \{(x, y, z, w) \ : \ xyzw = 0\}$,

(c) $q = 5$, $U = \{(\lambda + \mu, 2\mu, 3\lambda + v, v) \ : \ \lambda, \mu, v \in \mathbf{F}_5\}$,

(d) q prime, $V = \{(x_1, x_2, x_3) \ : \ x_1 = x_2 - x_3\}$.

4.3 For any given positive integer n and any $0 \le k \le n$, determine the number of distinct subspaces of $\mathbf{F}_q^n$ of dimension k.

4.4 (a) Let $\mathbf{F}_q$ be a subfield of $\mathbf{F}_r$. Show that $\mathbf{F}_r$ is a vector space over $\mathbf{F}_q$, where the vector addition and the scalar multiplication are the same as the addition and multiplication of the elements in the field $\mathbf{F}_r$, respectively.

(b) Let α be a root of an irreducible polynomial of degree m over $\mathbf{F}_q$. Show that $\{1, \alpha, \alpha^2, \ldots, \alpha^{m-1}\}$ is a basis of $\mathbf{F}_{q^m}$ over $\mathbf{F}_q$.

4.5 Define $\mathrm{Tr}_{\mathbf{F}_{q^m}/\mathbf{F}_q}(\alpha) = \alpha + \alpha^q + \cdots + \alpha^{q^{m-1}}$ for any $\alpha \in \mathbf{F}_{q^m}$. The element $\mathrm{Tr}_{\mathbf{F}_{q^m}/\mathbf{F}_q}(\alpha)$ is called the *trace* of α with respect to the extension $\mathbf{F}_{q^m}/\mathbf{F}_q$.

(i) Show that $\mathrm{Tr}_{\mathbf{F}_{q^m}/\mathbf{F}_q}(\alpha)$ is an element of $\mathbf{F}_q$ for all $\alpha \in \mathbf{F}_{q^m}$.

(ii) Show that the map

$$\mathrm{Tr}_{\mathbf{F}_{q^m}/\mathbf{F}_q} : \mathbf{F}_{q^m} \to \mathbf{F}_q, \quad \alpha \mapsto \mathrm{Tr}_{\mathbf{F}_{q^m}/\mathbf{F}_q}(\alpha)$$

is an $\mathbf{F}_q$-linear transformation, where both $\mathbf{F}_{q^m}$ and $\mathbf{F}_q$ are viewed as vector spaces over $\mathbf{F}_q$.

(iii) Show that $\mathrm{Tr}_{\mathbf{F}_{q^m}/\mathbf{F}_q}$ is surjective.

(iv) Let $\beta \in \mathbf{F}_{q^m}$. Prove that $\mathrm{Tr}_{\mathbf{F}_{q^m}/\mathbf{F}_q}(\beta) = 0$ if and only if there exists an element $\gamma \in \mathbf{F}_{q^m}$ such that $\beta = \gamma^q - \gamma$. (Note: this statement is commonly referred to as the additive form of Hilbert's Theorem 90.)

(v) (Transitivity of trace.) Prove that

$$\mathrm{Tr}_{\mathbf{F}_{q^{rm}}/\mathbf{F}_q}(\alpha) = \mathrm{Tr}_{\mathbf{F}_{q^m}/\mathbf{F}_q}(\mathrm{Tr}_{\mathbf{F}_{q^{rm}}/\mathbf{F}_{q^m}}(\alpha))$$

for any $\alpha \in \mathbf{F}_{q^{rm}}$.

4.6 (a) Let V be a vector space over a finite field $\mathbf{F}_q$. Show that $(\lambda \mathbf{u} + \mu \mathbf{v}) \cdot \mathbf{w} = \lambda(\mathbf{u} \cdot \mathbf{w}) + \mu(\mathbf{v} \cdot \mathbf{w})$, for all $\mathbf{u}, \mathbf{v}, \mathbf{w} \in V$ and $\lambda, \mu \in \mathbf{F}_q$.

(b) Give an example of a finite field $\mathbf{F}_q$ and a vector $\mathbf{u}$ defined over $\mathbf{F}_q$ with the property that $\mathbf{u} \ne \mathbf{0}$ but $\mathbf{u} \cdot \mathbf{u} = 0$.

(c) Let V be a vector space over a finite field $\mathbf{F}_q$ and let $\{\mathbf{v}_1, \mathbf{v}_2, \ldots, \mathbf{v}_k\}$ be a basis of V. Show that the following two statements are equivalent:

(i) $\mathbf{v} \cdot \mathbf{v}' = 0$ for all $\mathbf{v}, \mathbf{v}' \in V$,

(ii) $\mathbf{v}_i \cdot \mathbf{v}_j = 0$ for all $i, j \in \{1, 2, \ldots, k\}$.

(Note: this shows that it suffices to check (ii) when we need to determine whether a given linear code is self-orthogonal.)

4.7 Let $\mathbf{F}_q$ be a finite field and let S be a subset of $\mathbf{F}_q^n$.

(i) Show that $S^\perp$ and $<S>^\perp$ are subspaces of $\mathbf{F}_q^n$.

(ii) Show that $S^\perp = <S>^\perp$.

4.8 For each of the following sets S and corresponding finite fields $\mathbf{F}_q$, find the $\mathbf{F}_q$-linear span $<S>$ and its orthogonal complement $S^\perp$:

(a) $S = \{101, 111, 010\}, q = 2$,

(b) $S = \{1020, 0201, 2001\}, q = 3$,

(c) $S = \{00101, 10001, 11011\}, q = 2$.

Problems 4.9 to 4.13 deal with some well known inner products other than the Euclidean inner product.

4.9 Let $\langle, \rangle_H : \mathbf{F}_{q^2}^n \times \mathbf{F}_{q^2}^n \to \mathbf{F}_{q^2}$ be defined as

$$\langle \mathbf{u}, \mathbf{v} \rangle_H = \sum_{i=1}^n u_i v_i^q,$$

where $\mathbf{u} = (u_1, \ldots, u_n)$, $\mathbf{v} = (v_1, \ldots, v_n) \in \mathbf{F}_{q^2}^n$. Show that $\langle, \rangle_H$ is an inner product on $\mathbf{F}_{q^2}^n$. (Note: this inner product is called the *Hermitian inner product*. For a linear code C over $\mathbf{F}_{q^2}$, its Hermitian dual is defined as

$$C^{\perp_H} = \{\mathbf{v} \in \mathbf{F}_{q^2}^n : \langle \mathbf{v}, \mathbf{c} \rangle_H = 0 \text{ for all } \mathbf{c} \in C\}.$$

If $C = C^{\perp_H}$, then we say C is self-dual with respect to the Hermitian inner product.)

4.10 Write $\mathbf{F}_4 = \{0, 1, \alpha, \alpha^2\}$ (cf. Example 3.3.5). Show that the following linear codes over $\mathbf{F}_4$ are self-dual with respect to the Hermitian inner product:

(a) $C_1 = \{(0, 0), (1, 1), (\alpha, \alpha), (\alpha^2, \alpha^2)\}$;

(b) C_2 is the $\mathbf{F}_4$-linear code with generator matrix

$$\begin{pmatrix} 1 & 0 & 0 & 1 & \alpha & \alpha \\ 0 & 1 & 0 & \alpha & 1 & \alpha \\ 0 & 0 & 1 & \alpha & \alpha & 1 \end{pmatrix}.$$

(Note: the code C_2 is called the *hexacode*.)

Are C_1 and C_2 self-dual with respect to the Euclidean inner product?

4.11 Let $\langle,\rangle_S : \mathbf{F}_q^{2n} \times \mathbf{F}_q^{2n} \to \mathbf{F}_q$ be defined as

$$\langle (\mathbf{u}, \mathbf{v}), (\mathbf{u}', \mathbf{v}') \rangle_S = \mathbf{u} \cdot \mathbf{v}' - \mathbf{v} \cdot \mathbf{u}',$$

where $\mathbf{u}, \mathbf{v}, \mathbf{u}', \mathbf{v}' \in \mathbf{F}_q^n$ and $\cdot$ is the Euclidean inner product on $\mathbf{F}_q^n$. Show that $\langle,\rangle_S$ is an inner product on $\mathbf{F}_q^{2n}$. (Note: this inner product is called the *symplectic inner product*. It is useful in the construction of quantum error-correcting codes.)

4.12 For $(\mathbf{u}, \mathbf{v}) \in \mathbf{F}_q^{2n}$, where $\mathbf{u} = (u_1, \ldots, u_n)$ and $\mathbf{v} = (v_1, \ldots, v_n)$, the *symplectic weight* $\mathrm{wt}_S((\mathbf{u}, \mathbf{v}))$ of $(\mathbf{u}, \mathbf{v})$ is defined to be the number of $1 \le i \le n$ such that at least one of u_i, v_i is nonzero. Show that

$$\tfrac{1}{2}\mathrm{wt}((\mathbf{u}, \mathbf{v})) \le \mathrm{wt}_S((\mathbf{u}, \mathbf{v})) \le \mathrm{wt}((\mathbf{u}, \mathbf{v})),$$

where $\mathrm{wt}((\mathbf{u}, \mathbf{v}))$ denotes the usual Hamming weight of $(\mathbf{u}, \mathbf{v})$.

4.13 Let C be a linear code over $\mathbf{F}_q$ with a generator matrix $(I_n|A)$, where I_n is the $n \times n$ identity matrix and A is an $n \times n$ matrix satisfying $A = A^\mathrm{T}$.
  (i) Show that C is self-dual with respect to the symplectic inner product $\langle,\rangle_S$, i.e., $C = C^{\perp_S}$, where

$$C^{\perp_S} = \{\mathbf{v} \in \mathbf{F}_q^{2n} \ : \ \langle \mathbf{v}, \mathbf{c} \rangle_S = 0 \text{ for all } \mathbf{c} \in C\}.$$

  (ii) Show that C is equivalent to $C^\perp$, its dual under the usual Euclidean inner product.

4.14 Determine which of the following codes are linear over $\mathbf{F}_q$:
  (a) $q = 2$ and $C = \{1101, 1110, 1011, 1111\}$,
  (b) $q = 3$ and $C = \{0000, 1001, 0110, 2002, 1111, 0220, 1221, 2112, 2222\}$,
  (c) $q = 2$ and $C = \{00000, 11110, 01111, 10001\}$.

4.15 Let C and D be linear codes over $\mathbf{F}_q$ of the same length. Define

$$C + D = \{\mathbf{c} + \mathbf{d} \ : \ \mathbf{c} \in C, \mathbf{d} \in D\}.$$

Show that $C + D$ is a linear code and that $(C + D)^\perp = C^\perp \cap D^\perp$.

4.16 Determine whether each of the following statements is true or false. Justify your answer.
  (a) If C and D are linear codes over $\mathbf{F}_q$ of the same length, then $C \cap D$ is also a linear code over $\mathbf{F}_q$.
  (b) If C and D are linear codes over $\mathbf{F}_q$ of the same length, then $C \cup D$ is also a linear code over $\mathbf{F}_q$.
  (c) If $C = \langle S \rangle$, where $S = \{\mathbf{v}_1, \mathbf{v}_2, \mathbf{v}_3\} \subseteq \mathbf{F}_q^n$, then $\dim(C) = 3$.
  (d) If $C = \langle S \rangle$, where $S = \{\mathbf{v}_1, \mathbf{v}_2, \mathbf{v}_3\} \subseteq \mathbf{F}_q^n$, then

$$d(C) = \min\{\mathrm{wt}(\mathbf{v}_1), \mathrm{wt}(\mathbf{v}_2), \mathrm{wt}(\mathbf{v}_3)\}.$$

  (e) If C and D are linear codes over $\mathbf{F}_q$ with $C \subseteq D$, then $D^\perp \subseteq C^\perp$.

4.17 Determine the number of binary linear codes with parameters $[n, n-1, 2]$ for $n \geq 2$.

4.18 Prove Lemma 4.3.6.

4.19 Let $\mathbf{u} \in F_2^n$. A binary code C of length n is said to *correct the error pattern* $\mathbf{u}$ if and only if, for all $\mathbf{c}, \mathbf{c}' \in C$ with $\mathbf{c}' \neq \mathbf{c}$, we have $d(\mathbf{c}, \mathbf{c}+\mathbf{u}) < d(\mathbf{c}', \mathbf{c}+\mathbf{u})$. Assume that $\mathbf{u}_1, \mathbf{u}_2 \in F_2^n$ agree in at least the positions where 1 occurs in $\mathbf{u}_1$. Suppose that C corrects the error pattern $\mathbf{u}_2$. Prove that C also corrects the error pattern $\mathbf{u}_1$.

4.20 (i) Let $\mathbf{x}, \mathbf{y} \in F_2^n$. If $\mathbf{x}$ and $\mathbf{y}$ are both of even weight or both of odd weight, show that $\mathbf{x} + \mathbf{y}$ must have even weight.

(ii) Let $\mathbf{x}, \mathbf{y} \in F_2^n$. If exactly one of $\mathbf{x}, \mathbf{y}$ has even weight and the other has odd weight, show that $\mathbf{x} + \mathbf{y}$ must have odd weight.

(iii) Using (i) and (ii), or otherwise, prove that, for a binary linear code C, either all the codewords have even weight or exactly half of the codewords have even weight.

4.21 Let C be a binary linear code of parameters $[n, k, d]$. Assume that C has at least one codeword of odd weight. Let C' denote the subset of C consisting of all the codewords of even weight. Show that C' is a binary linear code of parameters $[n, k-1, d']$, with $d' > d$ if d is odd, and $d' = d$ if d is even. (Note: this is an example of an *expurgated* code.)

4.22 (a) Show that every codeword in a self-orthogonal binary code has even weight.

(b) Show that the weight of every codeword in a self-orthogonal ternary code is divisible by 3.

(c) Construct a self-orthogonal code over F_5 such that at least one of its codewords has weight *not* divisible by 5.

(d) Let $\mathbf{x}, \mathbf{y}$ be codewords in a self-orthogonal binary code. Suppose the weights of $\mathbf{x}$ and $\mathbf{y}$ are both divisible by 4. Show that the weight of $\mathbf{x} + \mathbf{y}$ is also a multiple of 4.

4.23 Let C be a self-dual binary code with parameters $[n, k, d]$.

(i) Show that the all-one vector $(1, 1, \ldots, 1)$ is in C.

(ii) Show that either all the codewords in C have weight divisible by 4; or exactly half of the codewords in C have weight divisible by 4 while the other half have even weight not divisible by 4.

(iii) Let $n = 6$. Determine d.

4.24 Give a parity-check matrix for a self-dual binary code of length 10.

4.25 Prove that there is no self-dual binary code of parameters $[10, 5, 4]$.

4.26 For n odd, let C be a self-orthogonal binary $[n, (n-1)/2]$-code. Let $\mathbf{1}$ denote the all-one vector of length n and let $\mathbf{1} + C = \{\mathbf{1} + \mathbf{c} : \mathbf{c} \in C\}$. Show that $C^\perp = C \cup (\mathbf{1} + C)$.

4.27 Let C be a linear code over $\mathbf{F}_q$ of length n. For any given i with $1 \le i \le n$, show that either the ith position of every codeword of C is 0 or every element $\alpha \in \mathbf{F}_q$ appears in the ith position of exactly $1/q$ of the codewords of C.

4.28 Let C be a linear code over $\mathbf{F}_q$ of parameters $[n, k, d]$ and suppose that, for every $1 \le i \le n$, there is at least one codeword whose ith position is nonzero.

(i) Show that the sum of the weights of all the codewords in C is $n(q - 1)q^{k-1}$.

(ii) Show that $d \le n(q - 1)q^{k-1}/(q^k - 1)$.

(iii) Show that there cannot be a binary linear code of parameters $[15, 7, d]$ with $d \ge 8$.

4.29 Let $\mathbf{x}, \mathbf{y}$ be two linearly independent vectors in $\mathbf{F}_q^n$ and let z denote the number of coordinates where $\mathbf{x}, \mathbf{y}$ are both 0.

(i) Show that $\mathrm{wt}(\mathbf{y}) + \sum_{\lambda \in \mathbf{F}_q} \mathrm{wt}(\mathbf{x} + \lambda\mathbf{y}) = q(n - z)$.

(ii) Suppose further that $\mathbf{x}, \mathbf{y}$ are contained in an $[n, k, d]$-code C over $\mathbf{F}_q$. Show that $\mathrm{wt}(\mathbf{x}) + \mathrm{wt}(\mathbf{y}) \le qn - (q - 1)d$.

4.30 Let C be an $[n, k, d]$-code over $\mathbf{F}_q$, where $\gcd(d, q) = 1$. Suppose that all the codewords of C have weight congruent to 0 or d modulo q.

(i) If $\mathbf{x}, \mathbf{y}$ are linearly independent codewords such that $\mathrm{wt}(\mathbf{x}) \equiv \mathrm{wt}(\mathbf{y}) \equiv 0 \pmod{q}$, show that $\mathrm{wt}(\mathbf{x} + \lambda\mathbf{y}) \equiv 0 \pmod{q}$ for all $\lambda \in \mathbf{F}_q$. (Hint: use Exercise 4.29.)

(ii) Show that $C_0 = \{\mathbf{c} \in C : \mathrm{wt}(\mathbf{c}) \equiv 0 \pmod{q}\}$ is a linear subcode of C; i.e., C_0 is a linear code contained in C.

(iii) Show that C cannot have a linear subcode of dimension 2 all of whose nonzero codewords have weight congruent to $d \pmod{q}$. Hence, deduce that C_0 has dimension $k - 1$.

(iv) Given a generator matrix G_0 for C_0 and a codeword $\mathbf{v} \in C$ of weight d, show that

$$\begin{pmatrix} \mathbf{v} \\ \hline G_0 \end{pmatrix}$$

is a generator matrix for C.

4.31 Find a generator matrix and a parity-check matrix for the linear code generated by each of the following sets, and give the parameters $[n, k, d]$ for each of these codes:

(a) $q = 2$, $S = \{1000, 0110, 0010, 0001, 1001\}$,

(b) $q = 3$, $S = \{110000, 011000, 001100, 000110, 000011\}$,

(c) $q = 2$, $S = \{10101010, 11001100, 11110000, 01100110, 00111100\}$.

4.32 Assign messages to the words in $\mathbf{F}_2^3$ as follows:

000	100	010	001	110	101	011	111
A	C	D	E	G	I	N	O

Let C be the binary linear code with generator matrix

$$G = \begin{pmatrix} 10101 \\ 01010 \\ 00011 \end{pmatrix}.$$

Use G to encode the message ENCODING.

4.33 Find a generator matrix G' in standard form for a binary linear code equivalent to the binary linear code with the given generator matrix G:

(a) $G = \begin{pmatrix} 1\,0\,1\,0\,1\,0 \\ 0\,1\,0\,1\,0\,1 \\ 1\,1\,0\,1\,1\,0 \\ 0\,0\,1\,0\,1\,1 \end{pmatrix}$,   (b) $G = \begin{pmatrix} 1\,0\,1\,1\,0\,0\,1\,1\,1 \\ 0\,0\,0\,1\,0\,1\,1\,0\,0 \\ 0\,0\,0\,1\,0\,1\,1\,1\,0 \end{pmatrix}.$

4.34 Find a generator matrix G' in standard form for a binary linear code C' equivalent to the binary linear code C with the given parity-check matrix H:

(a) $H = \begin{pmatrix} 1\,1\,0\,0\,0 \\ 0\,1\,1\,0\,1 \\ 0\,0\,0\,1\,1 \end{pmatrix}$,   (b) $H = \begin{pmatrix} 0\,1\,0\,1\,1\,1\,0 \\ 1\,1\,1\,1\,0\,0\,0 \\ 0\,1\,1\,0\,1\,0\,1 \end{pmatrix}.$

4.35 Construct a binary code C of length 6 as follows: for every $(x_1, x_2, x_3) \in \mathbf{F}_2^3$, construct a 6-bit word $(x_1, x_2, x_3, x_4, x_5, x_6) \in C$, where

$$x_4 = x_1 + x_2 + x_3,$$
$$x_5 = x_1 + x_3,$$
$$x_6 = x_2 + x_3.$$

(i) Show that C is a linear code.

(ii) Find a generator matrix and a parity-check matrix for C.

4.36 Construct a binary code C of length 8 as follows: for every $(a, b, c, d) \in \mathbf{F}_2^4$, construct an 8-bit word $(a, b, c, d, w, x, y, z) \in C$, where

$$w = a + b + c,$$
$$x = a + b + d,$$
$$y = a + c + d,$$
$$z = b + c + d.$$

(i) Show that C is a linear code.

(ii) Find a generator matrix and a parity-check matrix for C.

(iii) Show that C is exactly three-error-detecting and one-error-correcting.

(iv) Show that C is self-dual.

4.37 (a) Prove that equivalent linear codes always have the same length, dimension and distance.

(b) Show that, if C and C' are equivalent, then so are their duals $C^\perp$ and $(C')^\perp$.

4.38 Suppose that an $(n - k) \times n$ matrix H is a parity-check matrix for a linear code C over $\mathbf{F}_q$. Show that, if M is an invertible $(n - k) \times (n - k)$ matrix with entries in $\mathbf{F}_q$, then MH is also a parity-check matrix for C.

4.39 Find the distance of the binary linear code C with each of the following given parity-check matrices:

$$\text{(a) } H = \begin{pmatrix} 0111000 \\ 1110100 \\ 1100010 \\ 1010001 \end{pmatrix}, \quad \text{(b) } H = \begin{pmatrix} 1101000 \\ 1010100 \\ 0110010 \\ 1100001 \end{pmatrix}.$$

4.40 Let $n \geq 4$ and let H be a parity-check matrix for a binary linear code C of length n. Suppose that the columns of H are all distinct and that the weight of every column of H is odd. Show that the distance of C is at least 4.

4.41 List the cosets of each of the following q-ary linear codes:

(a) $q = 3$ and $C_3 = \{0000, 1010, 2020, 0101, 0202, 1111, 1212, 2121, 2222\}$,

(b) $q = 2$ and $C_2 = \{00000, 10001, 01010, 11011, 00100, 10101, 01110, 11111\}$.

4.42 Let H denote the parity-check matrix of a linear code C. Show that the coset of C whose syndrome is $\mathbf{v}$ contains a vector of weight t if and only if $\mathbf{v}$ is equal to some linear combination of t columns of H.

4.43 For m, n satisfying $2^{m-1} \leq n < 2^m$, let C be the binary $[n, n - m]$-code whose parity-check matrix H has as its ith column ($1 \leq i \leq n$) the binary representation of i (i.e., the first column is $(0 \ldots 01)^\mathrm{T}$, the second column is $(0 \ldots 010)^\mathrm{T}$ and the third column is $(0 \ldots 011)^\mathrm{T}$, etc.). Show that every coset of C contains a vector of weight ≤ 2.

4.44 Let $C \subseteq \mathbf{F}_q^n$ be a linear code with distance d. Show that a word $\mathbf{x} \in \mathbf{F}_q^n$ is the unique coset leader of $\mathbf{x} + C$ if $\mathrm{wt}(\mathbf{x}) \leq \lfloor (d - 1)/2 \rfloor$.

4.45 Let C be a linear code of distance d, where d is even. Show that some coset of C contains two vectors of weight $e + 1$, where $e = \lfloor (d - 1)/2 \rfloor$.

4.46 Show that

$$\begin{pmatrix} 1020 \\ 0102 \end{pmatrix}$$

is a parity-check matrix for C_3 in Exercise 4.41 and that

$$\begin{pmatrix} 10001 \\ 01010 \end{pmatrix}$$

is a parity-check matrix for C_2 in Exercise 4.41 . Using these parity-check matrices and assuming complete decoding, construct a syndrome look-up table for each of C_3 and C_2.

4.47 Let C be the binary linear code with parity-check matrix

$$H = \begin{pmatrix} 110100 \\ 101010 \\ 011001 \end{pmatrix}.$$

Write down a generator matrix for C and list all the codewords in C. Decode the following words:

(a) 110110,   (b) 011011,   (c) 101010.

4.48 Let p be a prime and let ζ denote a primitive pth root of unity in $\mathbf{C}$, the field of complex numbers (i.e., $\zeta^p = 1$ but $\zeta^i \neq 1$ for all $0 < i < p$). Let f be a function defined on $\mathbf{F}_p^n$ such that the values $f(\mathbf{v})$, where $\mathbf{v} \in \mathbf{F}_p^n$, can be added and subtracted, and multiplied naturally by complex numbers. Define the *discrete Fourier transform* $\hat{f}$ of f as follows:

$$\hat{f}(\mathbf{u}) = \sum_{\mathbf{v} \in \mathbf{F}_p^n} f(\mathbf{v}) \zeta^{\mathbf{u} \cdot \mathbf{v}},$$

where $\mathbf{u} \cdot \mathbf{v}$ is the Euclidean inner product in $\mathbf{F}_p^n$. Let C be a linear code of length n over $\mathbf{F}_p$ and, for $\mathbf{v} \in \mathbf{F}_p^n$, define

$$C_i(\mathbf{v}) = \{\mathbf{u} \in C : \mathbf{u} \cdot \mathbf{v} = i\}, \qquad \text{for } 0 \leq i \leq p - 1.$$

(i) Show that, for $1 \leq i \leq p - 1$, $C_i(\mathbf{v})$ is a coset of $C_0(\mathbf{v})$ in C if and only if $\mathbf{v} \notin C^\perp$. Show also that, if $\mathbf{v} \notin C^\perp$, then $C = C_0(\mathbf{v}) \cup C_1(\mathbf{v}) \cup \cdots \cup C_{p-1}(\mathbf{v})$.

(ii) Show that

$$\sum_{\mathbf{u} \in C} \zeta^{\mathbf{u} \cdot \mathbf{v}} = \begin{cases} |C| & \text{if } \mathbf{v} \in C^\perp, \\ 0 & \text{if } \mathbf{v} \notin C^\perp. \end{cases}$$

(iii) Show that $f(\mathbf{w}) = \dfrac{1}{p^n} \sum_{\mathbf{u} \in \mathbf{F}_p^n} \hat{f}(\mathbf{u}) \zeta^{-\mathbf{u} \cdot \mathbf{w}}$, where $\mathbf{w} \in \mathbf{F}_p^n$.

(iv) Show that $\displaystyle\sum_{\mathbf{v}\in C^\perp} f(\mathbf{v}) = \frac{1}{|C|}\sum_{\mathbf{u}\in C} \hat{f}(\mathbf{u})$.

4.49 Let C be a linear code of length n over $\mathbf{F}_p$, where p is a prime. The *(Hamming) weight enumerator* of C is the homogeneous polynomial

$$W_C(x, y) = \sum_{\mathbf{u}\in C} x^{n-\mathrm{wt}(\mathbf{u})} y^{\mathrm{wt}(\mathbf{u})}.$$

By setting $f(\mathbf{u}) = x^{n-\mathrm{wt}(\mathbf{u})} y^{\mathrm{wt}(\mathbf{u})}$ in Exercise 4.48, or otherwise, show that

$$W_{C^\perp}(x, y) = \frac{1}{|C|} W_C(x + (p-1)y, x - y).$$

(Note: this identity is called the *MacWilliams identity*. It actually holds for all finite fields $\mathbf{F}_q$, with p replaced by q in the above, though the proof is slightly more complicated.)

5 Bounds in coding theory

Given a q-ary (n, M, d)-code, where n is fixed, the size M is a measure of the efficiency of the code, and the distance d is an indication of its error-correcting capability. It would be nice if both M and d could be as large as possible, but, as we shall see shortly in this chapter, this is not quite possible, and a compromise needs to be struck.

For given q, n and d, we shall discuss some well known upper and lower bounds for the largest possible value of M. In the case where M is actually equal to one of the well known bounds, interesting codes such as perfect codes and MDS codes are obtained. We also discuss certain properties and examples of some of these fascinating families.

5.1 The main coding theory problem

Let C be a q-ary code with parameters (n, M, d). Recall from Chapter 2 that the information rate (or transmission rate) of C is defined to be $\mathcal{R}(C) = (\log_q M)/n$. We also introduce here the notion of the relative minimum distance.

Definition 5.1.1 For a q-ary code C with parameters (n, M, d), the *relative minimum distance* of C is defined to be $\delta(C) = (d - 1)/n$.

Remark 5.1.2 The relative minimum distance of C is often defined to be d/n in the literature, but defining it as $(d - 1)/n$ leads sometimes to neater formulas (see Remark 5.4.4).

Example 5.1.3 (i) Consider the q-ary code $C = \mathbf{F}_q^n$. It is easy to see that $(n, M, d) = (n, q^n, 1)$ or, alternatively, $[n, k, d] = [n, n, 1]$. Hence,

$$\mathcal{R}(C) = \frac{\log_q(q^n)}{n} = 1,$$

$$\delta(C) = 0.$$

This code has the maximum possible information rate, while its relative minimum distance is 0. As the minimum distance of a code is related closely to its error-correcting capability (cf. Theorem 2.5.10), a low relative minimum distance implies a relatively low error-correcting capability.

(ii) Consider the binary repetition code

$$C = \{\underbrace{00\cdots0}_{n}, \underbrace{11\cdots1}_{n}\}.$$

Clearly, $(n, M, d) = (n, 2, n)$ or, equivalently, C is a binary $[n, 1, n]$-linear code. Hence,

$$\mathcal{R}(C) = \frac{\log_2(2)}{n} = \frac{1}{n} \to 0,$$

$$\delta(C) = \frac{n-1}{n} \to 1,$$

as $n \to \infty$. As this code has the largest possible relative minimum distance, it has excellent error-correcting potential. However, this is achieved at the cost of very low efficiency, as reflected in the low information rate.

(iii) There is a family of binary linear codes (called Hamming codes – see Section 5.3.1) with parameters $(n, M, d) = (2^r - 1, 2^{n-r}, 3)$ or, equivalently, $[n, k, d] = [2^r - 1, 2^r - 1 - r, 3]$, for all integers $r \geq 2$. When $r \to \infty$, we have

$$\mathcal{R}(C) = \frac{\log_2(2^{n-r})}{n} = \frac{2^r - 1 - r}{2^r - 1} \to 1,$$

$$\delta(C) = \frac{2}{n} \to 0.$$

Again, while this family of codes has good information rates asymptotically, the relative minimum distances tend to zero, implying asymptotically bad error-correcting capabilities.

The previous examples should make it clear that a compromise between the transmission rate and the quality of error-correction is necessary.

Definition 5.1.4 For a given code alphabet A of size q (with $q > 1$) and given values of n and d, let $A_q(n, d)$ denote the largest possible size M for which there exists an (n, M, d)-code over A. Thus,

$$A_q(n, d) = \max\{M : \text{there exists an } (n, M, d)\text{-code over } A\}.$$

Any (n, M, d)-code C that has the maximum size, that is, for which $M = A_q(n, d)$, is called an *optimal code*.

Remark 5.1.5 (i) Note that $A_q(n, d)$ depends only on the size of A, n and d. It is independent of A.

(ii) The numbers $A_q(n, d)$ play a central role in coding theory, and much effort has been made in determining their values. In fact, the problem of determining the values of $A_q(n, d)$ is sometimes known as the *main coding theory problem*.

Instead of considering all codes, we may restrict ourselves to linear codes and obtain the following definition:

Definition 5.1.6 For a given prime power q and given values of n and d, let $B_q(n, d)$ denote the largest possible size q^k for which there exists an $[n, k, d]$-code over $\mathbf{F}_q$. Thus,

$$B_q(n, d) = \max\{q^k : \text{there exists an } [n, k, d]\text{-code over } \mathbf{F}_q\}.$$

While it is, in general, rather difficult to determine the exact values of $A_q(n, d)$ and $B_q(n, d)$, there are some properties that afford easy proofs.

Theorem 5.1.7 *Let $q \geq 2$ be a prime power. Then*

(i) $B_q(n, d) \leq A_q(n, d) \leq q^n$ *for all $1 \leq d \leq n$;*
(ii) $B_q(n, 1) = A_q(n, 1) = q^n$;
(iii) $B_q(n, n) = A_q(n, n) = q$.

Proof. The first inequality in (i) is obvious from the definitions, while the second one is clear since any (n, M, d)-code over $\mathbf{F}_q$, being a nonempty subset of $\mathbf{F}_q^n$, must have $M \leq q^n$.

To show (ii), note that $\mathbf{F}_q^n$ is an $[n, n, 1]$-linear code, and hence an $(n, q^n, 1)$-code, over $\mathbf{F}_q$, so $q^n \leq B_q(n, 1) \leq q^n$; i.e., $B_q(n, 1) = A_q(n, 1) = q^n$.

For (iii), let C be an (n, M, n)-code over $\mathbf{F}_q$. Since the codewords are of length n, and the distance between two distinct codewords is $\geq n$, it follows that the distance between two distinct codewords is actually n. This means that two distinct codewords must differ at all the coordinates. Therefore, at each coordinate, all the M words must take different values, so $M \leq q$, implying $B_q(n, n) \leq A_q(n, n) \leq q$. The repetition code of length n, i.e., $\{(a, a, \ldots, a) : a \in \mathbf{F}_q\}$, is an $[n, 1, n]$-linear code, and hence an (n, q, n)-code, over $\mathbf{F}_q$, so $B_q(n, n) = A_q(n, n) = q$. $\qquad\square$

In the case of binary codes, there are additional elementary results on $A_2(n, d)$ and $B_2(n, d)$. Before we discuss them, we need to introduce the notion of the extended code, which is a useful concept in its own right. For a binary linear code, its extended code is obtained by adding a parity-check coordinate. This idea can be generalized to codes over any finite field.

Definition 5.1.8 For any code C over $\mathbf{F}_q$, the *extended code of* C, denoted by $\overline{C}$, is defined to be

$$\overline{C} = \left\{ \left(c_1, \ldots, c_n, -\sum_{i=1}^{n} c_i \right) : (c_1, \ldots, c_n) \in C \right\}.$$

When $q = 2$, the extra coordinate $-\sum_{i=1}^{n} c_i = \sum_{i=1}^{n} c_i$ added to the codeword $(c_1, \ldots, c_n)$ is called the *parity-check* coordinate.

Theorem 5.1.9 *If C is an (n, M, d)-code over $\mathbf{F}_q$, then $\overline{C}$ is an $(n+1, M, d')$- code over $\mathbf{F}_q$, with $d \le d' \le d+1$. If C is linear, then so is $\overline{C}$. Moreover, when C is linear,*

$$\left(\begin{array}{c|c} H & \begin{array}{c} 0 \\ \vdots \\ 0 \end{array} \\ \hline 1 \cdots 1 & 1 \end{array} \right)$$

is a parity-check matrix of $\overline{C}$ if H is a parity-check matrix of C.

The proof is straightforward, so it is left to the reader (Exercise 5.3).

Example 5.1.10 (i) Consider the binary linear code $C_1 = \{000, 110, 011, 101\}$. It has parameters $[3, 2, 2]$. The extended code

$$\overline{C_1} = \{0000, 1100, 0110, 1010\}$$

is a binary $[4, 2, 2]$-linear code.

(ii) Consider the binary linear code $C_2 = \{000, 111, 011, 100\}$. It has parameters $[3, 2, 1]$. The extended code

$$\overline{C_2} = \{0000, 1111, 0110, 1001\}$$

is a binary $[4, 2, 2]$-linear code.

This example shows that the minimum distance $d(\overline{C})$ can achieve both $d(C)$ and $d(C) + 1$. Example 5.1.10(ii) is an illustration of the following fact.

Theorem 5.1.11 *Suppose d is odd.*

(i) *Then a binary (n, M, d)-code exists if and only if a binary $(n+1, M, d+1)$- code exists. Therefore, if d is odd, $A_2(n + 1, d + 1) = A_2(n, d)$.*

(ii) *Similarly, a binary* $[n, k, d]$-*linear code exists if and only if a binary* $[n+1,$
$k, d + 1]$-*linear code exists, so* $B_2(n + 1, d + 1) = B_2(n, d)$.

Proof. For (i), the latter statement follows immediately from the previous one, so we only prove the earlier statement.

Suppose that there is a binary (n, M, d)-code C, where d is odd. Then, from Theorem 5.1.9, $\overline{C}$ is an $(n + 1, M, d')$-code with $d \le d' \le d + 1$.

Note that $\mathrm{wt}(\mathbf{x}')$ is even for all $\mathbf{x}' \in \overline{C}$. Therefore, Lemma 4.3.5 and Corollary 4.3.4 show that $d(\mathbf{x}', \mathbf{y}')$ is even for all $\mathbf{x}', \mathbf{y}' \in \overline{C}$, so d' is even. Since d is odd and $d \le d' \le d + 1$, it follows that $d' = d + 1$.

We have therefore shown that, if there is a binary (n, M, d)-code C, then $\overline{C}$ is a binary $(n + 1, M, d + 1)$-code.

Next, we suppose that there exists a binary $(n + 1, M, d + 1)$-code D, where d is odd. Choose codewords $\mathbf{x}$ and $\mathbf{y}$ in D such that $d(\mathbf{x}, \mathbf{y}) = d + 1$. In other words, $\mathbf{x}$ and $\mathbf{y}$ differ at $d + 1 \ge 2$ coordinates. Choose a coordinate where $\mathbf{x}$ and $\mathbf{y}$ differ, and let D' be the code obtained by deleting this coordinate from all the codewords of D. (The code D' is called a *punctured code*; see Theorem 6.1.1(iii).) Then D' is a binary (n, M, d)-code.

For (ii), it suffices to observe that, in the proof of (i), if C is linear, then so is $\overline{C}$; similarly, if D is linear, then so is D'.   □

Remark 5.1.12 The last statement in Theorem 5.1.11(i) is equivalent to 'if d is even, then $A_2(n, d) = A_2(n - 1, d - 1)$'. There is also an analogue for (ii).

While the determination of the exact values of $A_q(n, d)$ and $B_q(n, d)$ can be rather difficult, several well known bounds, both upper and lower ones, do exist. We shall discuss some of them in the following sections.

A list of lower bounds and, in some cases, exact values for $A_2(n, d)$ may be found at the following webpage maintained by Simon Litsyn of Tel Aviv University:

http://www.eng.tau.ac.il/~litsyn/tableand/index.html.

The following website, maintained by Andries E. Brouwer of Technische Universiteit Eindhoven, contains tables that give the best known bounds (upper and lower) on the distance d for q-ary linear codes ($q \le 9$) of given length and dimension:

http://www.win.tue.nl/~aeb/voorlincod.html.

5.2 Lower bounds

We discuss two well known lower bounds: the sphere-covering bound (for $A_q(n, d)$) and the Gilbert–Varshamov bound (for $B_q(n, d)$).

5.2.1 Sphere-covering bound

Definition 5.2.1 Let A be an alphabet of size q, where $q > 1$. For any vector $\mathbf{u} \in A^n$ and any integer $r \geq 0$, the *sphere* of radius r and centre $\mathbf{u}$, denoted $S_A(\mathbf{u}, r)$, is the set $\{\mathbf{v} \in A^n \ : \ d(\mathbf{u}, \mathbf{v}) \leq r\}$.

Definition 5.2.2 For a given integer $q > 1$, a positive integer n and an integer $r \geq 0$, define $V_q^n(r)$ to be

$$V_q^n(r) = \begin{cases} \binom{n}{0} + \binom{n}{1}(q-1) + \binom{n}{2}(q-1)^2 + \cdots + \binom{n}{r}(q-1)^r & \text{if } 0 \leq r \leq n \\ q^n & \text{if } n \leq r. \end{cases}$$

Lemma 5.2.3 *For all integers $r \geq 0$, a sphere of radius r in A^n contains exactly $V_q^n(r)$ vectors, where A is an alphabet of size $q > 1$.*

Proof. Fix a vector $\mathbf{u} \in A^n$. We determine the number of vectors $\mathbf{v} \in A^n$ such that $d(\mathbf{u}, \mathbf{v}) = m$; i.e., the number of vectors in A^n of distance exactly m from $\mathbf{u}$. The number of ways in which to choose the m coordinates where $\mathbf{v}$ differs from $\mathbf{u}$ is given by $\binom{n}{m}$. For each coordinate, we have $q - 1$ choices for that coordinate in $\mathbf{v}$. Therefore, the total number of vectors of distance m from $\mathbf{u}$ is given by $\binom{n}{m}(q - 1)^m$. For $0 \leq r \leq n$, Lemma 5.2.3 now follows.
  When $r \geq n$, note that $S_A(\mathbf{u}, r) = A^n$, hence it contains $V_q^n(r) = q^n$ vectors.
$\square$

We are now ready to state and prove the sphere-covering bound.

Theorem 5.2.4 (Sphere-covering bound.) *For an integer $q > 1$ and integers n, d such that $1 \leq d \leq n$, we have*

$$\frac{q^n}{\sum_{i=0}^{d-1} \binom{n}{i}(q-1)^i} \leq A_q(n, d).$$

Proof. Let $C = \{\mathbf{c}_1, \mathbf{c}_2, \ldots, \mathbf{c}_M\}$ be an optimal (n, M, d)-code over A with $|A| = q$, so $M = A_q(n, d)$. Since C has the maximum size, there can be no word in A^n whose distance from every codeword in C is at least d. If there were such a word, we could simply include it in C, and thereby obtain an $(n, M + 1, d)$-code.

Therefore, for every vector $\mathbf{x}$ in A^n, there is at least one codeword $\mathbf{c}_i$ in C such that $d(\mathbf{x}, \mathbf{c}_i)$ is at most $d - 1$; i.e., $\mathbf{x} \in S_A(\mathbf{c}_i, d - 1)$. Hence, every word in A^n is contained in at least one of the spheres $S_A(\mathbf{c}_i, d - 1)$. In other words,

$$A^n \subseteq \bigcup_{i=1}^{M} S_A(\mathbf{c}_i, d - 1).$$

(For this reason, we say that the spheres $S_A(\mathbf{c}_i, d - 1)$ $(1 \le i \le M)$ cover A^n, hence the name 'sphere-covering' bound.)

Since $|A^n| = q^n$ and $|S_A(\mathbf{c}_i, d - 1)| = V_q^n(d - 1)$ for any i, we have

$$q^n \le M \cdot V_q^n(d - 1),$$

implying that

$$\frac{q^n}{V_q^n(d - 1)} \le M = A_q(n, d).$$

$\square$

Some examples of the lower bounds for $A_q(n, d)$ given by the sphere-covering bound are found in Tables 5.2–5.4 (see Example 5.5.5).

The following example illustrates how $A_q(n, d)$ may be found in some special cases. In the example, the lower bound is given by the sphere-covering bound. Then a combinatorial argument shows that the lower bound must also be an upper bound for $A_q(n, d)$, hence yielding the exact value of $A_q(n, d)$.

Example 5.2.5 We prove that $A_2(5, 4) = 2$.

The sphere-covering bound shows that $A_2(5, 4) \ge 2$.

By Theorem 5.1.11, we see that $A_2(5, 4) = A_2(4, 3)$, so we next show that $A_2(4, 3) \le 2$. Let C be a binary $(4, M, 3)$-code and let (x_1, x_2, x_3, x_4) be a codeword in C. Since $d(C) = 3$, the other codewords in C must be of the following forms:

$$(x_1, \overline{x_2}, \overline{x_3}, \overline{x_4}), \quad (\overline{x_1}, x_2, \overline{x_3}, \overline{x_4}), \quad (\overline{x_1}, \overline{x_2}, x_3, \overline{x_4}),$$
$$(\overline{x_1}, \overline{x_2}, \overline{x_3}, x_4), \quad (\overline{x_1}, \overline{x_2}, \overline{x_3}, \overline{x_4}),$$

where $\overline{x_i}$ is defined by

$$\overline{x_i} = \begin{cases} 1 & \text{if } x_i = 0 \\ 0 & \text{if } x_i = 1. \end{cases}$$

However, no pair of these five words are of distance 3 (or more) apart, and so only one of them can be included in C. Hence, $M \le 2$, implying that $A_2(4, 3) \le 2$. Therefore, $A_2(5, 4) = A_2(4, 3) = 2$.

5.2.2   Gilbert–Varshamov bound

The Gilbert–Varshamov bound is a lower bound for $B_q(n, d)$ (i.e., for linear codes) known since the 1950s. There is also an asymptotic version of the Gilbert–Varshamov bound, which concerns infinite sequences of codes whose lengths tend to infinity. However, we shall not discuss this asymptotic result here. The interested reader may refer to Chap. 17, Theorem 30 of ref. [13]. For a long time, the asymptotic Gilbert–Varshamov bound was the best lower bound known to be attainable by an infinite family of linear codes, so it became a sort of benchmark for judging the 'goodness' of an infinite sequence of linear codes. Between 1977 and 1982, V. D. Goppa constructed algebraic-geometry codes using algebraic curves over finite fields with many rational points. A major breakthrough in coding theory was achieved shortly after these discoveries, when it was shown that there are sequences of algebraic-geometry codes that perform better than the asymptotic Gilbert–Varshamov bound for certain sufficiently large q.

Theorem 5.2.6 (Gilbert–Varshamov bound.) *Let n, k and d be integers satisfying $2 \le d \le n$ and $1 \le k \le n$. If*

$$\sum_{i=0}^{d-2} \binom{n-1}{i}(q-1)^i < q^{n-k}, \tag{5.1}$$

then there exists an $[n, k]$-linear code over $\mathbf{F}_q$ with minimum distance at least d.

Proof. We shall show that, if (5.1) holds, then there exists an $(n-k) \times n$ matrix H over $\mathbf{F}_q$ such that every $d-1$ columns of H are linearly independent.

We construct H as follows. Let $\mathbf{c}_j$ denote the jth column of H.

Let $\mathbf{c}_1$ be any nonzero vector in $\mathbf{F}_q^{n-k}$. Let $\mathbf{c}_2$ be any vector not in the span of $\mathbf{c}_1$. For any $2 \le j \le n$, let $\mathbf{c}_j$ be any vector that is not in the linear span of $d-2$ (or fewer) of the vectors $\mathbf{c}_1, \ldots, \mathbf{c}_{j-1}$.

Note that the number of vectors in the linear span of $d-2$ or fewer of $\mathbf{c}_1, \ldots, \mathbf{c}_{j-1}$ ($2 \le j \le n$) is given by

$$\sum_{i=0}^{d-2} \binom{j-1}{i}(q-1)^i \le \sum_{i=0}^{d-2} \binom{n-1}{i}(q-1)^i < q^{n-k}.$$

Hence, the vector $\mathbf{c}_j$ ($2 \le j \le n$) can always be found.

The matrix H constructed in this manner is an $(n-k) \times n$ matrix, and any $d-1$ of its columns are linearly independent. The null space of H is a linear code over $\mathbf{F}_q$ of length n, of distance at least d, and of dimension at least k.

By turning to a k-dimensional subspace, we obtain a linear code of the desired type.   □

Corollary 5.2.7 *For a prime power $q > 1$ and integers n, d such that $2 \le d \le n$, we have*

$$B_q(n, d) \ge q^{n - \lceil \log_q (V_q^{n-1}(d-2)+1) \rceil} \ge \frac{q^{n-1}}{V_q^{n-1}(d - 2)}.$$

Proof. Put

$$k = n - \lceil \log_q (V_q^{n-1}(d - 2) + 1) \rceil.$$

Then (5.1) is satisfied and thus there exists a q-ary $[n, k, d_1]$-linear code with $d_1 \ge d$ by Theorem 5.2.6. By changing certain $d_1 - d$ fixed coordinates to 0, we obtain a q-ary $[n, k, d]$-linear code (see also Theorem 6.1.1(iv)). Our result follows from the fact that $B_q(n, d) \ge q^k$.   □

5.3   Hamming bound and perfect codes

The first upper bound for $A_q(n, d)$ that we will discuss is the Hamming bound, also known as the sphere-packing bound.

Theorem 5.3.1 (Hamming or sphere-packing bound.) *For an integer $q > 1$ and integers n, d such that $1 \le d \le n$, we have*

$$A_q(n, d) \le \frac{q^n}{\sum_{i=0}^{\lfloor (d-1)/2 \rfloor} \binom{n}{i}(q - 1)^i}.$$

Proof. Let $C = \{c_1, c_2, \ldots, c_M\}$ be an optimal (n, M, d)-code over A (with $|A| = q$), so $M = A_q(n, d)$. Let $e = \lfloor (d - 1)/2 \rfloor$; then the packing spheres $S_A(c_i, e)$ are disjoint. Hence, we have

$$\bigcup_{i=1}^{M} S_A(c_i, e) \subseteq A^n,$$

where the union on the left hand side is a disjoint union. Since $|A^n| = q^n$ and $|S_A(c_i, e)| = V_q^n(e)$ for any i, we have

$$M \cdot V_q^n(e) \le q^n,$$

implying that

$$A_q(n, d) = M \le \frac{q^n}{V_q^n(e)} = \frac{q^n}{V_q^n(\lfloor (d - 1)/2 \rfloor)}.$$

This completes the proof.   □

Definition 5.3.2 A q-ary code that attains the Hamming (or sphere-packing) bound, i.e., one which has $q^n / \left(\sum_{i=0}^{\lfloor (d-1)/2 \rfloor} \binom{n}{i} (q-1)^i \right)$ codewords, is called a *perfect* code.

Some of the earliest known codes, such as the Hamming codes and the Golay codes, are perfect codes.

5.3.1 Binary Hamming codes

Hamming codes were discovered by R. W. Hamming and M. J. E. Golay. They form an important class of codes – they have interesting properties and are easy to encode and decode.

While Hamming codes are defined over all finite fields $\mathbf{F}_q$, we begin by discussing specifically the binary Hamming codes. These codes form a special case of the general q-ary Hamming codes, but because they can be described more simply than the general q-ary Hamming codes, and because they are arguably the most interesting Hamming codes, it is worthwhile discussing them separately from the other Hamming codes.

Definition 5.3.3 Let $r \geq 2$. A binary linear code of length $n = 2^r - 1$, with parity-check matrix H whose columns consist of all the nonzero vectors of $\mathbf{F}_2^r$, is called a *binary Hamming code* of length $2^r - 1$. It is denoted by Ham(r, 2).

Remark 5.3.4 (i) The order of the columns of H has not been fixed in Definition 5.3.3. Hence, for each $r \geq 2$, the binary Hamming code Ham(r, 2) is only well defined up to equivalence of codes.

(ii) Note that the rows of H are linearly independent since H contains all the r columns of weight 1 words. Hence, H is indeed a parity-check matrix.

Example 5.3.5 Ham(3, 2): A Hamming code of length 7 with a parity-check matrix

$$H = \begin{pmatrix} 0 & 0 & 0 & 1 & 1 & 1 & 1 \\ 0 & 1 & 1 & 0 & 0 & 1 & 1 \\ 1 & 0 & 1 & 0 & 1 & 0 & 1 \end{pmatrix}.$$

Proposition 5.3.6 (Properties of the binary Hamming codes.)

  (i) *All the binary Hamming codes of a given length are equivalent.*
 (ii) *The dimension of* Ham(r, 2) *is* $k = 2^r - 1 - r$.
(iii) *The distance of* Ham(r, 2) *is* $d = 3$, *hence* Ham(r, 2) *is exactly single-error-correcting.*
(iv) *Binary Hamming codes are perfect codes.*

Proof. (i) For a given length, any parity-check matrix can be obtained from another by a permutation of the columns, hence the corresponding binary Hamming codes are equivalent.

(ii) Since H, a parity-check matrix for Ham$(r, 2)$, is an $r \times (2^r - 1)$ matrix, the dimension of Ham$(r, 2)$ is $2^r - 1 - r$.

(iii) Since no two columns of H are equal, any two columns of H are linearly independent. On the other hand, H contains the columns $(100\ldots0)^{\mathrm{T}}$, $(010\ldots0)^{\mathrm{T}}$ and $(110\ldots0)^{\mathrm{T}}$, which form a linearly dependent set. Hence, by Corollary 4.5.7, the distance of Ham$(r, 2)$ is equal to 3. It then follows from Theorem 2.5.10 that Ham$(r, 2)$ is single-error-correcting.

(iv) It can be verified easily that Ham$(r, 2)$ satisfies the Hamming bound and is hence a perfect code.   □

Decoding with a binary Hamming code

Since Ham$(r, 2)$ is perfect single-error-correcting, the coset leaders are precisely the 2^r $(= n + 1)$ vectors of length n of weight ≤ 1. Let $\mathbf{e}_j$ denote the vector with 1 in the jth coordinate and 0 elsewhere. Then the syndrome of $\mathbf{e}_j$ is just $\mathbf{e}_j H^{\mathrm{T}}$, i.e., the transpose of the jth column of H.

Hence, if the columns of H are arranged in the order of increasing binary numbers (i.e., the jth column of H is just the binary representation of j; see Exercise 4.43), the decoding is given by:

Step 1: When $\mathbf{w}$ is received, calculate its syndrome $S(\mathbf{w}) = \mathbf{w}H^{\mathrm{T}}$.

Step 2: If $S(\mathbf{w}) = \mathbf{0}$, assume $\mathbf{w}$ was the codeword sent.

Step 3: If $S(\mathbf{w}) \neq \mathbf{0}$, then $S(\mathbf{w})$ is the binary representation of j, for some $1 \leq j \leq 2^r - 1$. *Assuming a single error*, the word $\mathbf{e}_j$ gives the error, so we take the sent word to be $\mathbf{w} - \mathbf{e}_j$ (or, equivalently, $\mathbf{w} + \mathbf{e}_j$).

Example 5.3.7 We construct a syndrome look-up table for the Hamming code given in Example 5.3.5, and use it to decode $\mathbf{w} = 1001001$ (see Table 5.1).

The syndrome is $\mathbf{w}H^{\mathrm{T}} = 010$, which gives the coset leader $\mathbf{e}_2 = 0100000$. We can then decode $\mathbf{w}$ as $\mathbf{w} - \mathbf{e}_2 = \mathbf{w} + \mathbf{e}_2 = 1101001$.

Definition 5.3.8 The dual of the binary Hamming code Ham$(r, 2)$ is called a binary *simplex code*. It is sometimes denoted by $S(r, 2)$.

Some of the properties of the simplex codes are contained in Exercise 5.19.

Table 5.1.

Coset leader $\mathbf{u}$	Syndrome $S(\mathbf{u})$
0000000	000
1000000	001
0100000	010
0010000	011
0001000	100
0000100	101
0000010	110
0000001	111

Definition 5.3.9 The *extended binary Hamming code*, denoted $\overline{\mathrm{Ham}(r, 2)}$, is the code obtained from $\mathrm{Ham}(r, 2)$ by adding a parity-check coordinate.

Proposition 5.3.10 (Properties of the extended binary Hamming codes.)

(i) $\overline{\mathrm{Ham}(r, 2)}$ *is a binary* $[2^r, 2^r - 1 - r, 4]$*-linear code.*
(ii) *A parity-check matrix* $\overline{H}$ *for* $\overline{\mathrm{Ham}(r, 2)}$ *is*

$$
\overline{H} = \left(\begin{array}{c|c} H & \begin{array}{c} 0 \\ \vdots \\ 0 \end{array} \\ \hline 1 \cdots 1 & 1 \end{array} \right),
$$

where H *is a parity-check matrix for* $\mathrm{Ham}(r, 2)$.

Proposition 5.3.10 follows immediately from Theorem 5.1.9 and the proof of Theorem 5.1.11.

Remark 5.3.11 The rate of transmission for $\overline{\mathrm{Ham}(r, 2)}$ is slower than that of $\mathrm{Ham}(r, 2)$, but the extended code is better suited for incomplete decoding.

Example 5.3.12 Let $r = 3$ and take

$$
\overline{H} = \begin{pmatrix} 0 & 0 & 0 & 1 & 1 & 1 & 1 & 0 \\ 0 & 1 & 1 & 0 & 0 & 1 & 1 & 0 \\ 1 & 0 & 1 & 0 & 1 & 0 & 1 & 0 \\ 1 & 1 & 1 & 1 & 1 & 1 & 1 & 1 \end{pmatrix}.
$$

Note that every codeword is made up of 8 bits and recall that the syndrome of the error vector $\mathbf{e}_j$ is just the transpose of the jth column of $\overline{H}$. Assuming that as few errors as possible have occurred, the incomplete decoding works as follows. Suppose the received vector is $\mathbf{w}$, so its syndrome is $S(\mathbf{w}) = \mathbf{w}\overline{H}^{\mathrm{T}}$. Suppose it is $S(\mathbf{w}) = (s_1, s_2, s_3, s_4)$. Then $S(\mathbf{w})$ must fall into one of the following four categories:

(i) $s_4 = 0$ and $(s_1, s_2, s_3) = \mathbf{0}$. In this case, $S(\mathbf{w}) = \mathbf{0}$, so $\mathbf{w} \in \overline{\mathrm{Ham}(3, 2)}$. We may therefore assume that there are no errors.

(ii) $s_4 = 0$ and $(s_1, s_2, s_3) \neq \mathbf{0}$. Since $S(\mathbf{w}) \neq \mathbf{0}$, at least one error must have occurred. If exactly one error occurs and it occurs in the jth bit, then the error vector is $\mathbf{e}_j$, so $S(\mathbf{w}) = S(\mathbf{e}_j)$, which is the transpose of the jth column of $\overline{H}$. An inspection of $\overline{H}$ shows immediately that the last coordinate (the one corresponding to s_4) of every column is 1, contradicting the fact that $s_4 = 0$. Hence, the assumption that exactly one error has occurred is flawed, and we may assume at least two errors have occurred and seek retransmission.

(iii) $s_4 = 1$ and $(s_1, s_2, s_3) = \mathbf{0}$. Again, since $S(\mathbf{w}) \neq \mathbf{0}$, at least one error has occurred. It is easy to see that $S(\mathbf{w}) = S(\mathbf{e}_8)$, so we may assume a single error in the last coordinate, i.e., the parity-check coordinate.

(iv) $s_4 = 1$ and $(s_1, s_2, s_3) \neq \mathbf{0}$. As before, it is easy to check that $S(\mathbf{w})$ must coincide with the transpose of one of the first seven columns of $\overline{H}$, say the jth column. Hence, $S(\mathbf{w}) = S(\mathbf{e}_j)$, and we may assume a single error in the jth coordinate. In fact, given the way the columns of $\overline{H}$ are arranged, j is the number whose binary representation is (s_1, s_2, s_3).

5.3.2 q-ary Hamming codes

Let $q \geq 2$ be any prime power. Note that any nonzero vector $\mathbf{v} \in \mathbf{F}_q^r$ generates a subspace $< \mathbf{v} >$ of dimension 1. Furthermore, for $\mathbf{v}, \mathbf{w} \in \mathbf{F}_q^r \backslash \{\mathbf{0}\}$, $< \mathbf{v} >=$ $< \mathbf{w} >$ if and only if there is a nonzero scalar $\lambda \in \mathbf{F}_q \backslash \{0\}$ such that $\mathbf{v} = \lambda \mathbf{w}$. Therefore, there are exactly $(q^r - 1)/(q - 1)$ distinct subspaces of dimension 1 in $\mathbf{F}_q^r$.

Definition 5.3.13 Let $r \geq 2$. A q-ary linear code, whose parity-check matrix H has the property that the columns of H are made up of precisely one nonzero vector from each vector subspace of dimension 1 of $\mathbf{F}_q^r$, is called a *q-ary Hamming code*, often denoted as $\mathrm{Ham}(r, q)$.

It is an easy exercise to show that, when $q = 2$, the code defined here is the same as the binary Hamming code defined earlier.

Remark 5.3.14 An easy way to write down a parity-check matrix for Ham(r, q) is to list as columns all the nonzero r-tuples in $\mathbf{F}_q^r$ whose first nonzero entry is 1.

Proposition 5.3.15 (Properties of the q-ary Hamming codes.)

(i) Ham(r, q) *is a* $[(q^r - 1)/(q - 1), (q^r - 1)/(q - 1) - r, 3]$-*code.*
(ii) Ham(r, q) *is a perfect exactly single-error-correcting code.*

The proof of Proposition 5.3.15 resembles that of Proposition 5.3.6, so we leave it as an exercise to the reader (Exercise 5.17).

Decoding with a q-ary Hamming code

Since Ham(r, q) is a perfect single-error-correcting code, the coset leaders, other than $\mathbf{0}$, are exactly the vectors of weight 1. A typical coset leader is then denoted by $\mathbf{e}_{j,b}$ ($1 \leq j \leq n$, $b \in \mathbf{F}_q \backslash \{0\}$) – the vector whose jth coordinate is b and the other coordinates are 0. Note that

$$S(\mathbf{e}_{j,b}) = b\mathbf{c}_j^\mathrm{T},$$

where $\mathbf{c}_j$ denotes the jth column of H.
Decoding works as follows:

Step 1: Given a received word $\mathbf{w}$, calculate $S(\mathbf{w}) = \mathbf{w}H^\mathrm{T}$.

Step 2: If $S(\mathbf{w}) = \mathbf{0}$, then assume no errors.

Step 3: If $S(\mathbf{w}) \neq \mathbf{0}$, then find the unique $\mathbf{e}_{j,b}$ such that $S(\mathbf{w}) = S(\mathbf{e}_{j,b})$. The received word is then taken to be $\mathbf{w} - \mathbf{e}_{j,b}$.

Definition 5.3.16 The dual of the q-ary Hamming code Ham(r, q) is called a q-ary *simplex code*. It is sometimes denoted by $S(r, q)$.

The reader may refer to Exercise 5.19 for some of the properties of the q-ary simplex codes.

5.3.3 Golay codes

The Golay codes were discovered by M. J. E. Golay in the late 1940s. The (unextended) Golay codes are examples of perfect codes. It turns out that the Golay codes are essentially unique in the sense that binary or ternary codes with the same parameters as them can be shown to be equivalent to them.

Binary Golay codes

Definition 5.3.17 Let G be the 12×24 matrix

$$G = (I_{12}|A),$$

where I_{12} is the 12×12 identity matrix and A is the 12×12 matrix

$$A = \begin{pmatrix}
0 & 1 & 1 & 1 & 1 & 1 & 1 & 1 & 1 & 1 & 1 & 1 \\
1 & 1 & 1 & 0 & 1 & 1 & 1 & 0 & 0 & 0 & 1 & 0 \\
1 & 1 & 0 & 1 & 1 & 1 & 0 & 0 & 0 & 1 & 0 & 1 \\
1 & 0 & 1 & 1 & 1 & 0 & 0 & 0 & 1 & 0 & 1 & 1 \\
1 & 1 & 1 & 1 & 0 & 0 & 0 & 1 & 0 & 1 & 1 & 0 \\
1 & 1 & 1 & 0 & 0 & 0 & 1 & 0 & 1 & 1 & 0 & 1 \\
1 & 1 & 0 & 0 & 0 & 1 & 0 & 1 & 1 & 0 & 1 & 1 \\
1 & 0 & 0 & 0 & 1 & 0 & 1 & 1 & 0 & 1 & 1 & 1 \\
1 & 0 & 0 & 1 & 0 & 1 & 1 & 0 & 1 & 1 & 1 & 0 \\
1 & 0 & 1 & 0 & 1 & 1 & 0 & 1 & 1 & 1 & 0 & 0 \\
1 & 1 & 0 & 1 & 1 & 0 & 1 & 1 & 1 & 0 & 0 & 0 \\
1 & 0 & 1 & 1 & 0 & 1 & 1 & 1 & 0 & 0 & 0 & 1
\end{pmatrix}.$$

The binary linear code with generator matrix G is called the *extended binary Golay code* and will be denoted by G_{24}.

Remark 5.3.18 (i) The Voyager 1 and 2 spacecraft were launched towards Jupiter and Saturn in 1977. This code was used in the encoding and decoding of the general science and engineering (GSE) data for the missions.

(ii) It is also common to call any code that is equivalent to the linear code with generator matrix G an extended binary Golay code.

Proposition 5.3.19 (Properties of the extended binary Golay code.)

(i) *The length of G_{24} is 24 and its dimension is 12.*

(ii) *A parity-check matrix for G_{24} is the 12×24 matrix*

$$H = (A|I_{12}).$$

(iii) *The code G_{24} is self-dual, i.e., $G_{24}^{\perp} = G_{24}$.*

(iv) *Another parity-check matrix for G_{24} is the 12×24 matrix*

$$H' = (I_{12}|A)(=G).$$

(v) *Another generator matrix for G_{24} is the 12×24 matrix*

$$G' = (A|I_{12})(=H).$$

(vi) *The weight of every codeword in G_{24} is a multiple of 4.*

(vii) *The code G_{24} has no codeword of weight 4, so the distance of G_{24} is $d = 8$.*

(viii) *The code G_{24} is an exactly three-error-correcting code.*

Proof. (i) This is clear from the definition.

(ii) This follows from Theorem 4.5.9.

(iii) Note that the rows of G are orthogonal; i.e., if $\mathbf{r}_i$ and $\mathbf{r}_j$ are any two rows of G, then $\mathbf{r}_i \cdot \mathbf{r}_j = 0$. This implies that $G_{24} \subseteq G_{24}^\perp$. On the other hand, since both G_{24} and $G_{24}^\perp$ have dimension 12, we must have $G_{24} = G_{24}^\perp$.

(iv) A parity-check matrix of G_{24} is a generator matrix of $G_{24}^\perp = G_{24}$, and G is one such matrix.

(v) A generator matrix of G_{24} is a parity-check matrix of $G_{24}^\perp = G_{24}$, and H is one such matrix.

(vi) Let $\mathbf{v}$ be a codeword in G_{24}. We want to show that wt($\mathbf{v}$) is a multiple of 4. Note that $\mathbf{v}$ is a linear combination of the rows of G. Let $\mathbf{r}_i$ denote the ith row of G.

First, suppose $\mathbf{v}$ is one of the rows of G. Since the rows of G have weight 8 or 12, the weight of $\mathbf{v}$ is a multiple of 4.

Next, let $\mathbf{v}$ be the sum $\mathbf{v} = \mathbf{r}_i + \mathbf{r}_j$ of two different rows of G. Since G_{24} is self-dual, Exercise 4.22(d) shows that the weight of $\mathbf{v}$ is divisible by 4.

We then continue by induction to finish the proof.

(vii) Note that the last row of G is a codeword of weight 8. This fact, together with statement (vi) of this proposition, implies that $d = 4$ or 8.

Suppose G_{24} contains a nonzero codeword $\mathbf{v}$ with wt($\mathbf{v}$) = 4. Write $\mathbf{v}$ as $(\mathbf{v}_1, \mathbf{v}_2)$, where $\mathbf{v}_1$ is the vector (of length 12) made up of the first 12 coordinates of $\mathbf{v}$, and $\mathbf{v}_2$ is the vector (also of length 12) made up of the last 12 coordinates of $\mathbf{v}$. Then one of the following situations must occur:

Case (1) wt($\mathbf{v}_1$) = 0 and wt($\mathbf{v}_2$) = 4. This cannot possibly happen since, by looking at the generator matrix G, the only such word is $\mathbf{0}$, which is of weight 0.

Case (2) wt($\mathbf{v}_1$) = 1 and wt($\mathbf{v}_2$) = 3. In this case, again by looking at G, $\mathbf{v}$ must be one of the rows of G, which is again a contradiction.

Case (3) wt($\mathbf{v}_1$) = 2 and wt($\mathbf{v}_2$) = 2. Then $\mathbf{v}$ is the sum of two of the rows of G. It is easy to check that none of such sums would give wt($\mathbf{v}_2$) = 2.

Case (4) wt($\mathbf{v}_1$) = 3 and wt($\mathbf{v}_2$) = 1. Since G' is a generator matrix, $\mathbf{v}$ must be one of the rows of G', which clearly gives a contradiction.

Case (5) wt($\mathbf{v}_1$) = 4 and wt($\mathbf{v}_2$) = 0. This case is similar to case (1), using G' instead of G.

Since we obtain contradictions in all these cases, $d = 4$ is impossible. Thus, $d = 8$.

(viii) This follows from statement (vii) above and Theorem 2.5.10.   □

Definition 5.3.20 Let $\hat{G}$ be the 12×23 matrix

$$\hat{G} = (I_{12}|\hat{A}),$$

where I_{12} is the 12×12 identity matrix and $\hat{A}$ is the 12×11 matrix obtained from the matrix A by deleting the last column of A. The binary linear code with generator matrix $\hat{G}$ is called the *binary Golay code* and will be denoted by G_{23}.

Remark 5.3.21 Alternatively, the binary Golay code can be defined as the code obtained from G_{24} by deleting the last coordinate of every codeword.

Proposition 5.3.22 (Properties of the binary Golay code.)

(i) *The length of G_{23} is 23 and its dimension is 12.*
(ii) *A parity-check matrix for G_{23} is the 11×23 matrix*

$$\hat{H} = (\hat{A}^T|I_{11}).$$

(iii) *The extended code of G_{23} is G_{24}.*
(iv) *The distance of G_{23} is $d = 7$.*
(v) *The code G_{23} is a perfect exactly three-error-correcting code.*

The proof is left as an exercise to the reader (see Exercise 5.24).

Ternary Golay codes

Definition 5.3.23 The *extended ternary Golay code*, denoted by G_{12}, is the ternary linear code with generator matrix $G = (I_6|B)$, where B is the 6×6 matrix

$$B = \begin{pmatrix} 0 & 1 & 1 & 1 & 1 & 1 \\ 1 & 0 & 1 & 2 & 2 & 1 \\ 1 & 1 & 0 & 1 & 2 & 2 \\ 1 & 2 & 1 & 0 & 1 & 2 \\ 1 & 2 & 2 & 1 & 0 & 1 \\ 1 & 1 & 2 & 2 & 1 & 0 \end{pmatrix}.$$

Remark 5.3.24 Any linear code that is equivalent to the above code is also called an extended ternary Golay code.

By mimicking the method used in Proposition 5.3.19, it is possible to check that G_{12} is a self-dual ternary $[12, 6, 6]$-code (see Exercise 5.28).

Definition 5.3.25 The *ternary Golay code* G_{11} is the code obtained by puncturing G_{12} in the last coordinate.

One can verify that G_{11} satisfies the Hamming bound and is hence a perfect ternary $[11, 6, 5]$-code (see Exercise 5.29).

5.3.4 Some remarks on perfect codes

The following codes are obviously perfect codes and are called *trivial perfect codes*:

(i)   the linear code $C = \mathbf{F}_q^n$ $(d = 1)$;
(ii) any C with $|C| = 1$ $(d = \infty)$;
(iii) binary repetition codes of odd lengths consisting of two codewords at distance n from each other $(d = n)$.

In the earlier subsections, we have seen that the Hamming codes and the Golay codes are examples of nontrivial perfect codes. Various constructions of nonlinear perfect codes with the same parameters as the q-ary Hamming codes have also been found.

In fact, the following result is true.

Theorem 5.3.26 (Van Lint and Tietäväinen.) *When $q \geq 2$ is a prime power, a nontrivial perfect code over $\mathbf{F}_q$ must have the same parameters as one of the Hamming or Golay codes.*

This result was obtained by Tietäväinen [22, 23] with considerable contribution from van Lint [12]. A proof may be found in Chap. 6 of ref. [13]. This result was also independently proved by Zinov'ev and Leont'ev [25].

5.4 Singleton bound and MDS codes

In this section, we discuss an upper bound for $A_q(n, d)$ due to Singleton [20].

Theorem 5.4.1 (Singleton bound.) *For any integer $q > 1$, any positive integer n and any integer d such that $1 \leq d \leq n$, we have*

$$A_q(n, d) \leq q^{n-d+1}.$$

In particular, when q is a prime power, the parameters $[n, k, d]$ of any linear code over $\mathbf{F}_q$ satisfy

$$k + d \leq n + 1.$$

Proof. We first note that the final statement of Theorem 5.4.1 follows from the previous one since, by definition of $A_q(n, d)$, $q^k \leq A_q(n, d)$.

To prove that $A_q(n, d) \leq q^{n-d+1}$, consider an (n, M, d)-code C over an alphabet A of size q, where $M = A_q(n, d)$. Delete the last $d - 1$ coordinates from all the codewords of C. Since the distance of C is d, after deleting the last $d - 1$ coordinates from all the codewords, the remaining words (of length $n - d + 1$) are still all distinct. The maximum number of words of length $n - d + 1$ is q^{n-d+1}, so $A_q(n, d) = M \leq q^{n-d+1}$. $\quad\square$

Remark 5.4.2 The following is another easy direct proof for the inequality $k + d \leq n + 1$ in the case of an $[n, k, d]$-linear code C:

Given any parity-check matrix H for C, the row rank, and hence the rank, of H is, by definition, $n - k$. Therefore, any $n - k + 1$ columns of H form a linearly dependent set. By Theorem 4.5.6(ii), $d \leq n - k + 1$.

Definition 5.4.3 A linear code with parameters $[n, k, d]$ such that $k + d = n + 1$ is called a *maximum distance separable (MDS)* code.

Remark 5.4.4 An alternative way to state the Singleton bound is: for any q-ary code C, we have

$$\mathcal{R}(C) + \delta(C) \leq 1.$$

(In this situation, we see that our choice of the definition of the relative minimum distance $\delta(C)$ gives a neater inequality than if $\delta(C)$ is defined to be d/n.) A linear code C is MDS if and only if $\mathcal{R}(C) + \delta(C) = 1$.

One of the interesting properties of MDS codes is the following.

Theorem 5.4.5 *Let C be a linear code over $\mathbf{F}_q$ with parameters $[n, k, d]$. Let G, H be a generator matrix and a parity-check matrix, respectively, for C. Then, the following statements are equivalent:*

  (i) *C is an MDS code;*
 (ii) *every set of $n - k$ columns of H is linearly independent;*
(iii) *every set of k columns of G is linearly independent;*
 (iv) *$C^{\perp}$ is an MDS code.*

Proof. The equivalence of (i) and (ii) follows directly from Corollary 4.5.7, with $d = n - k + 1$.

Since G is a parity-check matrix for $C^\perp$, (iii) and (iv) are also equivalent by Corollary 4.5.7.

Next, we prove that (i) implies (iv).

Recall that H is a generator matrix for $C^\perp$, so the length of $C^\perp$ is n and the dimension is $n - k$. To show that $C^\perp$ is MDS, we need to show that the minimum distance d' is $k + 1$.

Suppose $d' \leq k$. Then there is a word $\mathbf{c} \in C^\perp$ with at most k nonzero entries (and hence at least $n - k$ zero coordinates). Permuting the coordinates does not change the weight of the words, so we may assume that the last $n - k$ coordinates of $\mathbf{c}$ are 0.

Write H as $H = (A|H')$, where A is some $(n - k) \times k$ matrix and H' is a square $(n - k) \times (n - k)$ matrix. Since the columns of H' are linearly independent (for (i) and (ii) are equivalent), H' is invertible. Hence, the rows of H' are linearly independent. The only way to obtain 0 in all the last $n - k$ coordinates (such as for $\mathbf{c}$) is to use the 0-linear combination of the rows of H' (by linear independence). Therefore, the entire word $\mathbf{c}$ is the all-zero word $\mathbf{0}$. Consequently, $d' \geq k + 1$. Together with the Singleton bound, it now follows that $d' = k + 1$.

Since $(C^\perp)^\perp = C$, the above also shows that (iv) implies (i). This completes the proof of the theorem.   □

Definition 5.4.6 An MDS code C over $\mathbf{F}_q$ is *trivial* if and only if C satisfies one of the following:

(i) $C = \mathbf{F}_q^n$;
(ii) C is equivalent to the code generated by $\mathbf{1} = (1, \ldots, 1)$; or
(iii) C is equivalent to the dual of the code generated by $\mathbf{1}$.

Otherwise, C is said to be *nontrivial*.

Remark 5.4.7 When $q = 2$, the only MDS codes are the trivial ones. This fact follows easily by considering the generator matrix in standard form (see Exercise 5.32).

An interesting family of examples of MDS codes is given by the (generalized) Reed–Solomon codes. For more details, see Chapters 8 and 9. Some other examples may also be found in the exercises at the end of this chapter.

5.5 Plotkin bound

The next upper bound for $A_q(n, d)$ that we will discuss is the Plotkin bound, which holds for codes for which d is large relative to n. It often gives a tighter upper bound than many of the other upper bounds, though it is only applicable to a comparatively smaller range of values of d. The proof we give for the Plotkin bound makes use of the following well known Cauchy–Schwarz inequality.

Lemma 5.5.1 (Cauchy–Schwarz inequality.) *Let* $\{a_1, \ldots, a_m\}$ *and* $\{b_1, \ldots, b_m\}$ *be any two sets of real numbers. Then*

$$\left(\sum_{r=1}^{m} a_r b_r\right)^2 = \left(\sum_{r=1}^{m} a_r^2\right)\left(\sum_{s=1}^{m} b_s^2\right) - \sum_{r=1}^{m}\sum_{s=1}^{m}(a_r b_s - a_s b_r)^2/2$$

Consequently,

$$\left(\sum_{r=1}^{m} a_r b_r\right)^2 \leq \left(\sum_{r=1}^{m} a_r^2\right)\left(\sum_{r=1}^{m} b_r^2\right).$$

For more details on the Cauchy–Schwarz inequality, see, for example, ref. [9].

Theorem 5.5.2 (Plotkin bound.) *Let* $q > 1$ *be an integer and suppose that* n, d *satisfy* $rn < d$, *where* $r = 1 - q^{-1}$. *Then,*

$$A_q(n, d) \leq \left\lfloor \frac{d}{d - rn} \right\rfloor.$$

Proof. Let C be an (n, M, d)-code over an alphabet A of size q. Let

$$T = \sum_{\mathbf{c} \in C} \sum_{\mathbf{c}' \in C} d(\mathbf{c}, \mathbf{c}').$$

Since $d \leq d(\mathbf{c}, \mathbf{c}')$ for $\mathbf{c}, \mathbf{c}' \in C$ such that $\mathbf{c} \neq \mathbf{c}'$, it follows that

$$M(M - 1)d \leq T. \tag{5.2}$$

Now let $\mathcal{A}$ be the $M \times n$ array whose rows are made up of the M codewords in C. For $1 \leq i \leq n$ and $a \in A$, let $n_{i,a}$ denote the number of entries in the ith column of $\mathcal{A}$ that are equal to a. Hence, $\sum_{a \in A} n_{i,a} = M$ for every $1 \leq i \leq n$. Consequently, writing $\mathbf{c} = (c_1, \ldots, c_n)$ and $\mathbf{c}' = (c_1', \ldots, c_n')$, we have

$$T = \sum_{i=1}^{n}\left(\sum_{\mathbf{c} \in C}\sum_{\mathbf{c}' \in C} d(c_i, c_i')\right) = \sum_{i=1}^{n}\sum_{a \in A} n_{i,a}(M - n_{i,a}) = M^2 n - \sum_{i=1}^{n}\sum_{a \in A} n_{i,a}^2.$$

Applying Lemma 5.5.1, with $m = q$ and $a_1 = \cdots = a_q = 1$, it follows that

$$T \leq M^2 n - \sum_{i=1}^{n} q^{-1} \left(\sum_{a \in A} n_{i,a} \right)^2 = M^2 rn. \qquad (5.3)$$

The Plotkin bound now follows from (5.2) and (5.3). □

In fact, when $q = 2$, a more refined version of the Plotkin bound is available.

Theorem 5.5.3 (Plotkin bound for binary codes.)
(i) *When d is even,*

$$A_2(n, d) \leq \begin{cases} 2\lfloor d/(2d - n) \rfloor & \text{for } n < 2d \\ 4d & \text{for } n = 2d. \end{cases}$$

(ii) *When d is odd,*

$$A_2(n, d) \leq \begin{cases} 2\lfloor (d + 1)/(2d + 1 - n) \rfloor & \text{for } n < 2d + 1 \\ 4d + 4 & \text{for } n = 2d + 1. \end{cases}$$

We leave the proof of Theorem 5.5.3 as an exercise (Exercise 5.30).

Example 5.5.4 To illustrate that Theorem 5.5.3 gives a more refined bound than Theorem 5.5.2, note that Theorem 5.5.2 gives $A_2(8, 5) \leq 5$, $A_2(8, 6) \leq 3$, $A_2(12, 7) \leq 7$ and $A_2(11, 8) \leq 3$, whereas Theorem 5.5.3 gives $A_2(8, 5) \leq 4$, $A_2(8, 6) \leq 2$, $A_2(12, 7) \leq 4$ and $A_2(11, 8) \leq 2$.

Example 5.5.5 In Tables 5.2–5.4, we list the sphere-covering lower bound and compare the Hamming, Singleton and Plotkin upper bounds for $A_2(n, d)$, with $d = 3, 5, 7$ and $d \leq n \leq 12$. In cases where the Plotkin bound is not applicable, the entry is marked '–'.

5.6 Nonlinear codes

Whereas most of this book focuses on linear codes, there are several families of (binary) nonlinear codes that are well known and important in coding theory. We provide a brief introduction to some of them in this section.

Table 5.2. Bounds for $A_2(n, 3)$.

n	Sphere-covering	Hamming	Singleton	Plotkin
3	2	2	2	2
4	2	3	4	2
5	2	5	8	4
6	3	9	16	8
7	5	16	32	16
8	7	28	64	–
9	12	51	128	–
10	19	93	256	–
11	31	170	512	–
12	52	315	1024	–

Table 5.3. Bounds for $A_2(n, 5)$.

n	Sphere-covering	Hamming	Singleton	Plotkin
5	2	2	2	2
6	2	2	4	2
7	2	4	8	2
8	2	6	16	4
9	2	11	32	6
10	3	18	64	12
11	4	30	128	24
12	6	51	256	–

Table 5.4. Bounds for $A_2(n, 7)$.

n	Sphere-covering	Hamming	Singleton	Plotkin
7	2	2	2	2
8	2	2	4	2
9	2	3	8	2
10	2	5	16	2
11	2	8	32	4
12	2	13	64	4

5.6.1 Hadamard matrix codes

Definition 5.6.1 A *Hadamard matrix* H_n is an $n \times n$ integer matrix whose entries are 1 or -1 and which satisfies $H_n H_n^{\mathrm{T}} = n I_n$, where I_n is the identity matrix.

When such a Hadamard matrix exists, then either $n = 1, 2$ or n is a multiple of 4. The existence of Hadamard matrices is known for many n; for example when n is a power of 2 (these are called Sylvester matrices), and when $n = p^m + 1$, where p is a prime and n is divisible by 4 (this is called the Paley construction). The construction of Sylvester matrices is easy. We begin with $H_1 = (1)$ and use the observation that, whenever H_n is a Hadamard matrix of order n, the matrix

$$H_{2n} = \begin{pmatrix} H_n & H_n \\ H_n & -H_n \end{pmatrix}$$

is a Hadamard matrix of order $2n$.

The existence of a Hadamard matrix H_n implies the existence of binary nonlinear codes of the following parameters:

$$
\begin{array}{ll}
(n, 2\lfloor d/(2d - n) \rfloor, d) & \text{for } d \text{ even and } d \leq n < 2d; \\
(2d, 4d, d) & \text{for } d \text{ even}; \\
(n, 2\lfloor (d + 1)/(2d + 1 - n) \rfloor, d) & \text{for } d \text{ odd and } d \leq n < 2d + 1; \\
(2d + 1, 4d + 4, d) & \text{for } d \text{ odd}.
\end{array}
$$

These codes were constructed by Levenshtein [10]. By the Plotkin bound, they are optimal.

5.6.2 Nordstrom–Robinson code

It can be shown that there cannot be any binary linear codes of parameters $[16, 8, 6]$ (see Exercise 5.35). However, there does exist a binary nonlinear code, called the *Nordstrom–Robinson code*, of parameters $(16, 2^8, 6)$. It was discovered by Nordstrom and Robinson [16] (when Nordstrom was still a high school student!) and later independently by Semakov and Zinov'ev [19]. One construction of this famous code is as follows.

Rearrange the columns of the extended binary Golay code so that the new code (also called G_{24}) contains the word $111111110 \cdots 0$, and let G denote a generator matrix for this new G_{24}. Since $d(G_{24}) = 8 > 7$, Theorem 4.5.6(i) shows that the first seven columns of G are linearly independent. One can then show that each of the 2^7 possible vectors in $\mathbf{F}_2^7$ appears as the first seven coordinates of some codeword in G_{24}. In fact, each of them appears in exactly

$2^{12}/2^7 = 32$ codewords of G_{24}. Now collect all those words in G_{24} whose first seven coordinates are either all 0 or are made up of six 0s and one 1. There are altogether $8 \times 32 = 256 = 2^8$ of them.

The Nordstrom–Robinson code is obtained by deleting the first eight coordinates from these 2^8 vectors. It can be shown that this code has minimum distance 6 and is nonlinear.

5.6.3 Preparata codes

For $m \geq 2$, Preparata codes are binary nonlinear codes with the parameters $(2^{2m}, 2^{2^{2m}-4m}, 6)$.

There are several different ways to construct the Preparata codes; one way is as follows.

Write the vectors of $\mathbf{F}_2^{2^{2m}}$ in the form $(\mathbf{u}, \mathbf{v})$, where $\mathbf{u}, \mathbf{v} \in \mathbf{F}_2^{2^{2m-1}}$. Label the coordinate positions of these vectors in $\mathbf{F}_2^{2^{2m-1}}$ by the elements of $\mathbf{F}_{2^{2m-1}}$, with the first coordinate position corresponding to 0. For $\alpha \in \mathbf{F}_{2^{2m-1}}$, denote the entry at the αth coordinate of $\mathbf{u}, \mathbf{v}$ by u_α, v_α, respectively.

Definition 5.6.2 For $m \geq 2$, the *Preparata code* $P(m)$ of length 2^{2m} consists of all the codewords $(\mathbf{u}, \mathbf{v})$, where $\mathbf{u}, \mathbf{v} \in \mathbf{F}_2^{2^{2m-1}}$, satisfying the following conditions:

(i) both $\mathbf{u}$ and $\mathbf{v}$ are of even Hamming weight;

(ii) $\sum_{u_\alpha=1} \alpha = \sum_{v_\alpha=1} \alpha$;

(iii) $\sum_{u_\alpha=1} \alpha^3 + \left(\sum_{u_\alpha=1} \alpha \right)^3 = \sum_{v_\alpha=1} \alpha^3$.

It can be shown that $P(m)$ is a subcode of the extended binary Hamming code of the same length (see Chap. 15 of ref. [13] or Sect. 9.4 of ref. [24]).

The first code in this family, with $m = 2$, can be shown to be equivalent to the Nordstrom–Robinson $(16, 2^8, 6)$-code in Section 5.6.2.

5.6.4 Kerdock codes

For $m \geq 2$, the *Kerdock codes* $K(m)$ are binary nonlinear codes with parameters $(2^{2m}, 2^{4m}, 2^{2m-1} - 2^{m-1})$.

The Kerdock code $K(m)$ is constructed as a union of 2^{2m-1} cosets of the Reed–Muller code $\mathcal{R}(1, 2m)$ in $\mathcal{R}(2, 2m)$ (see Section 6.2).

Once again, the first code in this family, with $m = 2$, is equivalent to the Nordstrom–Robinson code. The Kerdock codes form a special case of a more general family of nonlinear codes called the *Delsarte–Goethals codes*.

The weight enumerators of the Kerdock and Preparata codes can be shown to satisfy the MacWilliams identity (see Exercise 4.49), thus giving a 'formal duality' between the Kerdock and Preparata codes. However, this falls beyond the scope of this book, so we will not elaborate further on this formal duality. The interested reader may refer to Chap. 15, Theorem 24 of ref. [13] for more details. This mystery of the formal duality between the Kerdock and Preparata codes was explained when it was shown by Nechaev [15] and Hammons *et al.* [7] that the Kerdock codes can be viewed as linear codes over the ring $\mathbf{Z}_4$, and by Hammons *et al.* [7] that the binary images of the $\mathbf{Z}_4$-dual of the Kerdock codes over $\mathbf{Z}_4$ can be regarded as variants of the Preparata codes.

5.7   Griesmer bound

The next bound we shall discuss is the Griesmer bound, which applies specifically to linear codes.

Let C be a linear code over $\mathbf{F}_q$ with parameters $[n, k]$ and suppose $\mathbf{c}$ is a codeword in C with $\mathrm{wt}(\mathbf{c}) = w$.

Definition 5.7.1 The *support* of $\mathbf{c}$, denoted by $\mathrm{Supp}(\mathbf{c})$, is the set of coordinates at which $\mathbf{c}$ is nonzero.

Definition 5.7.2 The *residual code of C with respect to $\mathbf{c}$*, denoted $\mathrm{Res}(C, \mathbf{c})$, is the code of length $n - w$ obtained from C by puncturing on all the coordinates of $\mathrm{Supp}(\mathbf{c})$.

Note that $w = |\mathrm{Supp}(\mathbf{c})|$.

Lemma 5.7.3 *If C is an $[n, k, d]$-code over $\mathbf{F}_q$ and $\mathbf{c} \in C$ is a codeword of weight d, then $\mathrm{Res}(C, \mathbf{c})$ is an $[n - d, k - 1, d']$-code, where $d' \geq \lceil d/q \rceil$. Here, $\lceil x \rceil$ is the least integer greater than or equal to x.*

Proof. Without loss of generality, we may replace C by an equivalent code so that $\mathbf{c} = (1, 1, \ldots, 1, 0, 0, \ldots, 0)$, where the first d coordinates are 1 and the other coordinates are 0.

We first note that $\mathrm{Res}(C, \mathbf{c})$ has dimension at most $k - 1$. To see this, observe first that $\mathrm{Res}(C, \mathbf{c})$ is a linear code. For every $\mathbf{x} \in \mathbf{F}_q^n$, denote by $\mathbf{x}'$ the vector obtained from $\mathbf{x}$ by deleting the first d coordinates, i.e., by puncturing on the coordinates of $\mathrm{Supp}(\mathbf{c})$. Now, it is easy to see that the map $C \rightarrow \mathrm{Res}(C, \mathbf{c})$ given by $\mathbf{x} \mapsto \mathbf{x}'$ is a well defined surjective linear transformation of vector spaces, whose kernel contains $\mathbf{c}$ and is hence a subspace of C of dimension at least 1. Therefore, $\mathrm{Res}(C, \mathbf{c})$ has dimension at most $k - 1$.

We shall show that $\mathrm{Res}(C, \mathbf{c})$ has dimension exactly $k - 1$.

Suppose that the dimension is strictly less than $k - 1$. Then there is a nonzero codeword $\mathbf{v} = (v_1, v_2, \ldots, v_n)$ in C that is not a multiple of $\mathbf{c}$ and that has the property that $v_{d+1} = \cdots = v_n = 0$. Then $\mathbf{v} - v_1\mathbf{c}$ is a nonzero codeword that belongs to C and that has weight strictly less than d, contradicting the definition of d. Hence, $\mathrm{Res}(C, \mathbf{c})$ has dimension $k - 1$.

To show that $d' \geq \lceil d/q \rceil$, let $(x_{d+1}, \ldots, x_n)$ be any nonzero codeword of $\mathrm{Res}(C, \mathbf{c})$, and let $\mathbf{x} = (x_1, \ldots, x_d, x_{d+1}, \ldots, x_n)$ be a corresponding word in C. By the pigeonhole principle, there is an $\alpha \in \mathbf{F}_q$ such that at least d/q coordinates of $(x_1, \ldots, x_d)$ are equal to α. Hence,

$$d \leq \mathrm{wt}(\mathbf{x} - \alpha\mathbf{c}) \leq d - \frac{d}{q} + \mathrm{wt}((x_{d+1}, \ldots, x_n)).$$

The inequality $d' \geq \lceil d/q \rceil$ now follows.   □

Theorem 5.7.4 (Griesmer bound.) *Let C be a q-ary code of parameters $[n, k, d]$, where $k \geq 1$. Then*

$$n \geq \sum_{i=0}^{k-1} \left\lceil \frac{d}{q^i} \right\rceil.$$

Proof. We prove the Griesmer bound by induction on k. Clearly, when $k = 1$, Theorem 5.7.4 holds.

When $k > 1$ and $\mathbf{c} \in C$ is a codeword of minimum weight d, then Lemma 5.7.3 shows that $\mathrm{Res}(C, \mathbf{c})$ is an $[n - d, k - 1, d']$-code, where $d' \geq \lceil d/q \rceil$. By the inductive hypothesis, we may assume that the Griesmer bound holds for $\mathrm{Res}(C, \mathbf{c})$, hence

$$n - d \geq \sum_{i=0}^{k-2} \left\lceil \frac{d'}{q^i} \right\rceil \geq \sum_{i=0}^{k-2} \left\lceil \frac{d}{q^{i+1}} \right\rceil.$$

Theorem 5.7.4 now follows.   □

Example 5.7.5 From Exercise 5.19, the q-ary simplex code $S(r, q)$ has parameters $[(q^r - 1)/(q - 1), r, q^{r-1}]$, so it meets the Griesmer bound.

5.8 Linear programming bound

One of the best bounds in existence for $A_q(n, d)$ is one that is based on linear programming techniques. It is due to Delsarte [2]. The bound obtained through this method is often called the linear programming bound.

A family of polynomials, called the *Krawtchouk polynomials*, plays a pivotal role in this theory. Krawtchouk polynomials are also very useful in other areas of coding theory. We give the definition and summarize some properties of these polynomials below.

Definition 5.8.1 For a given q, the *Krawtchouk polynomial* $K_k(x; n)$ is defined to be

$$K_k(x; n) = \sum_{j=0}^{k} (-1)^j \binom{x}{j} \binom{n-x}{k-j} (q-1)^{k-j}.$$

When there is no ambiguity for n, the notation is often simplified to $K_k(x)$.

Proposition 5.8.2 (Properties of Krawtchouk polynomials.)

(i) *If z is a variable, then $\sum_{k=0}^{\infty} K_k(x)z^k = (1 + (q-1)z)^{n-x}(1-z)^x$.*

(ii) $K_k(x) = \sum_{j=0}^{k} (-1)^j q^{k-j} \binom{n-k+j}{j} \binom{n-x}{k-j}$.

(iii) *$K_k(x)$ is a polynomial of degree k, with leading coefficient $(-q)^k/k!$ and constant term $K_k(0) = \binom{n}{k}(q-1)^k$.*

(iv) *(Orthogonality relations.)* $\sum_{i=0}^{n} \binom{n}{i}(q-1)^i K_k(i)K_\ell(i) = \delta_{k\ell}\binom{n}{k}$ $(q-1)^k q^n$, *where $\delta_{k\ell}$ is the Kronecker delta function; i.e.,*

$$\delta_{k\ell} = \begin{cases} 1 & \text{if } k = \ell \\ 0 & \text{otherwise.} \end{cases}$$

(v) $(q-1)^i \binom{n}{i} K_k(i) = (q-1)^k \binom{n}{k} K_i(k)$.

(vi) $\sum_{i=0}^{n} K_\ell(i)K_i(k) = \delta_{k\ell} q^n$.

(vii) $\sum_{k=0}^{j} \binom{n-k}{n-j} K_k(x) = q^j \binom{n-x}{j}$.

(viii) *When $q = 2$, we have $K_i(x)K_j(x) = \sum_{k=0}^{n} \binom{n-k}{(i+j-k)/2} \binom{k}{(i-j+k)/2} K_k(x)$.*

(ix) *Every polynomial $f(x)$ of degree r can be expressed as $f(x) = \sum_{k=0}^{r} f_k K_k(x)$, where $f_k = q^{-n} \sum_{i=0}^{n} f(i)K_i(k)$. (This way of expressing $f(x)$ is called the Krawtchouk expansion of $f(x)$.)*

We leave the proof of Proposition 5.8.2 to the reader (see Exercise 5.42).

The linear programming bound gives an upper bound for $A_q(n, d)$; i.e., it applies also to nonlinear codes. Therefore, we will deal with the distance between two distinct codewords and not the weight of each codeword. For the main result in this section, we need the following notion.

Definition 5.8.3 Let A be an alphabet of size q. For C an (n, M)-code over A and for all $0 \le i \le n$, let

$$A_i(C) = \frac{1}{M}|\{(\mathbf{u}, \mathbf{v}) \in C \times C \ : \ d(\mathbf{u}, \mathbf{v}) = i\}|.$$

The sequence $\{A_i(C)\}_{i=0}^{n}$ is called the *distance distribution* of C.

Remark 5.8.4 Note that the distance distribution depends only on the size q of the code alphabet and not on the alphabet itself. To obtain the linear programming bound, it is more convenient to work with the ring $\mathbf{Z}_q$ as the alphabet. Hence, in the discussion below, while we begin with codes over an alphabet A of size q, we pass immediately to codes over $\mathbf{Z}_q$ in the proofs.

Lemma 5.8.5 *Let C be a q-ary code of length n. Then*

$$\sum_{i=0}^{n} A_i(C)K_k(i) \ge 0$$

for all integers $0 \le k \le n$.

Proof. As mentioned in Remark 5.8.4, we assume C is defined over $\mathbf{Z}_q$. It suffices to show that $M \sum_{i=0}^{n} A_i(C)K_k(i) \ge 0$, where $M = |C|$. Using Exercise 5.46,

$$M \sum_{i=0}^{n} A_i(C)K_k(i) = \sum_{i=0}^{n} \sum_{\substack{(\mathbf{u},\mathbf{v})\in C^2 \\ d(\mathbf{u},\mathbf{v})=i}} \sum_{\substack{\mathbf{w}\in\mathbf{Z}_q^n \\ \mathrm{wt}(\mathbf{w})=k}} \zeta^{(\mathbf{u}-\mathbf{v})\cdot\mathbf{w}} = \sum_{\substack{\mathbf{w}\in\mathbf{Z}_q^n \\ \mathrm{wt}(\mathbf{w})=k}} \left| \sum_{\mathbf{u}\in C} \zeta^{\mathbf{u}\cdot\mathbf{w}} \right|^2 \ge 0,$$

where, for $\mathbf{u} = (u_1, \dots, u_n)$ and $\mathbf{w} = (w_1, \dots, w_n)$, $\mathbf{u} \cdot \mathbf{w} = u_1 w_1 + \cdots + u_n w_n$, and ζ is a primitive qth root of unity in $\mathbf{C}$; i.e., $\zeta^q = 1$ but $\zeta^i \ne 1$ for all $0 < i < q$. $\qquad\square$

Theorem 5.8.6 (Linear programming bound – version 1.) *For a given integer $q > 1$ and positive integers n and d $(1 \le d \le n)$, we have*

$$A_q(n, d) \le \max \left\{ \sum_{i=0}^{n} A_i \ : \ A_0 = 1, \ A_i = 0 \ \text{for } 1 \le i < d, \ A_i \ge 0 \text{ for } 0 \le i \le n \right.$$

$$\left. \sum_{i=0}^{n} A_i K_k(i) \ge 0 \text{ for } 0 \le k \le n \right\}. \tag{5.4}$$

Proof. Let $M = A_q(n, d)$. If C is a q-ary (n, M)-code, its distance distribution $\{A_i(C)\}_{i=0}^{n}$ satisfies the following conditions:

(i) $A_0(C) = 1$;

(ii) $A_i(C) = 0$ for $1 \leq i < d$;

(iii) $A_i(C) \geq 0$ for all $0 \leq i \leq n$;

(iv) $\sum_{i=0}^{n} A_i(C) K_k(i) \geq 0$ for $0 \leq k \leq n$ (from Lemma 5.8.5);

(v) $M = A_q(n, d) = \sum_{i=0}^{n} A_i(C)$.

Hence, the inequality (5.4) follows immediately.   □

The following theorem is the duality theorem of Theorem 5.8.6 in linear programming. It is often more useful than Theorem 5.8.6 because any polynomial $f(x)$ that satisfies Theorem 5.8.7 gives an upper bound for $A_q(n, d)$, while an optimal solution for the linear programming problem in (5.4) is required to give an upper bound for $A_q(n, d)$.

Theorem 5.8.7 (Linear programming bound – version 2.) *Let $q > 1$ be an integer. For positive integers n and d ($1 \leq d \leq n$), let $f(x) = 1 + \sum_{k=1}^{n} f_k K_k(x)$ be a polynomial such that $f_k \geq 0$ ($1 \leq k \leq n$) and $f(i) \leq 0$ for $d \leq i \leq n$. Then $A_q(n, d) \leq f(0)$.*

Proof. As in the proof of Theorem 5.8.6, let $M = A_q(n, d)$, let C be a q-ary (n, M)-code and let $\{A_i(C)\}_{i=0}^{n}$ be its distance distribution.

Note that conditions (i), (ii) and (iv) in the proof of Theorem 5.8.6 imply that $K_k(0) \geq -\sum_{i=d}^{n} A_i(C) K_k(i)$ for all $0 \leq k \leq n$. The condition that $f(i) \leq 0$ for $d \leq i \leq n$ implies that $\sum_{i=d}^{n} A_i(C) f(i) \leq 0$, which means that

$$f(0) = 1 + \sum_{k=1}^{n} f_k K_k(0)$$

$$\geq 1 - \sum_{k=1}^{n} f_k \sum_{i=d}^{n} A_i(C) K_k(i)$$

$$= 1 - \sum_{i=d}^{n} A_i(C) \sum_{k=1}^{n} f_k K_k(i)$$

$$= 1 - \sum_{i=d}^{n} A_i(C)(f(i) - 1)$$

$$\geq 1 + \sum_{i=d}^{n} A_i(C)$$

$$= M = A_q(n, d).$$

□

To illustrate that the linear programming bound can be better than some other bounds that we have discussed in this chapter, we show in Example 5.8.8

how one can deduce the Singleton bound, the Hamming bound and the Plotkin bound from the linear programming bound.

Example 5.8.8 (i) (Singleton bound.) Let

$$f(x) = q^{n-d+1} \prod_{j=d}^{n} \left(1 - \frac{x}{j}\right).$$

By Proposition 5.8.2(ix), $f(x) = \sum_{k=0}^{n} f_k K_k(x)$, where f_k is given by

$$
\begin{aligned}
f_k &= \frac{1}{q^n} \sum_{i=0}^{n} f(i) K_i(k) \\
&= \frac{1}{q^{d-1}} \sum_{i=0}^{d-1} \binom{n-i}{n-d+1} K_i(k) \Big/ \binom{n}{d-1} \\
&= \binom{n-k}{d-1} \Big/ \binom{n}{d-1} \geq 0,
\end{aligned}
$$

where the last equality follows from Proposition 5.8.2(vii). In particular, $f_0 = 1$. Clearly, $f(i) = 0$ for $d \leq i \leq n$.

Hence, by Theorem 5.8.7, it follows that $A_q(n, d) \leq f(0) = q^{n-d+1}$, which is the Singleton bound (cf. Theorem 5.4.1).

(ii) (Hamming bound.) Let $d = 2e + 1$. Let $f(x) = \sum_{k=0}^{n} f_k K_k(x)$, where

$$f_k = \left\{ L_e(k) \Big/ \sum_{i=0}^{e} (q-1)^i \binom{n}{i} \right\}^2 \qquad (0 \leq k \leq n),$$

with $L_e(x) = \sum_{i=0}^{e} K_i(x) = K_e(x-1; n-1)$. (The polynomial $L_e(x)$ is called a *Lloyd polynomial*.) Clearly, $f_k \geq 0$ for all $0 \leq k \leq n$ and $f_0 = 1$. Using Proposition 5.8.2(viii) and (vi), it can be shown that $f(i) = 0$ for $d \leq i \leq n$. Therefore, Theorem 5.8.7 and Proposition 5.8.2(iv) show that

$$A_q(n, d) \leq f(0) = q^n \Big/ \sum_{i=0}^{e} (q-1)^i \binom{n}{i},$$

which is exactly the Hamming bound.

(iii) (Plotkin bound for $A_2(2\ell + 1, \ell + 1)$.) Set $q = 2$, $n = 2\ell + 1$ and $d = \ell + 1$. Take $f_1 = (\ell + 1)/(2\ell + 1)$ and $f_2 = 1/(2\ell + 1)$, so that

$$
\begin{aligned}
f(x) &= 1 + \frac{\ell+1}{2\ell+1} K_1(x) + \frac{1}{2\ell+1} K_2(x) \\
&= 1 + \frac{\ell+1}{2\ell+1}(2\ell+1-2x) + \frac{1}{2\ell+1}(2x^2 - 2(2\ell+1)x + \ell(2\ell+1)).
\end{aligned}
$$

Clearly, $f_k \geq 0$ for all $1 \leq k \leq n$, and it is straightforward to verify that $f(i) \leq 0$ for $\ell + 1 = d \leq i \leq n = 2\ell + 1$. (In fact, $f(x)$ is a quadratic polynomial such that $f(\ell + 1) = 0 = f(2\ell + 1)$.)

Hence, by Theorem 5.8.7, it follows that

$$A_2(2\ell + 1, \ell + 1) \leq f(0) = 1 + \frac{\ell + 1}{2\ell + 1}(2\ell + 1) + \frac{1}{2\ell + 1}\ell(2\ell + 1) = 2\ell + 2,$$

which is exactly the Plotkin bound (cf. Theorem 5.5.2). (Note: when ℓ is even, Theorem 5.5.3 in fact gives a better bound.)

Exercises

5.1 Find the size, (minimum) distance, information rate and relative minimum distance of each of the following codes:
 (a) the binary code of all the words of length 3;
 (b) the ternary code consisting of all the words of length 4 whose second and fourth coordinates are 0;
 (c) the code over the alphabet $\mathbf{F}_p$ (p prime) consisting of all the words of length 3 whose first coordinate is $p - 1$ and whose second coordinate is 1;
 (d) the repetition code over the alphabet $\mathbf{F}_p$ (p prime) consisting of the following words of length n: $(0, 0, \ldots, 0), (1, 1, \ldots, 1), \ldots, (p - 1, p - 1, \ldots, p - 1)$.

5.2 For n odd, let C be a self-orthogonal binary $[n, (n - 1)/2]$-code. Show that $\overline{C^{\perp}}$ is a self-dual code. (Note: compare with Exercise 4.26.)

5.3 For any code C over $\mathbf{F}_q$ and any $\epsilon \in \mathbf{F}_q^*$, let

$$\overline{C}_\epsilon = \left\{ \left(c_1, \ldots, c_n, \epsilon \sum_{i=1}^{n} c_i \right) : (c_1, \ldots, c_n) \in C \right\}.$$

(In particular, $\overline{C}_{-1}$ is the extended code $\overline{C}$ of C defined in Definition 5.1.8.)
 (i) If C is an (n, M, d)-code, show that $\overline{C}_\epsilon$ is an $(n + 1, M, d')$-code, where $d \leq d' \leq d + 1$.
 (ii) If C is linear, show that $\overline{C}_\epsilon$ is linear also. Find a parity-check matrix for $\overline{C}_\epsilon$ in terms of a parity-check matrix H of C.

5.4 Without using any of the bounds discussed in this chapter, show that
 (a) $A_2(6, 5) = 2$,   (b) $A_2(7, 5) = 2$.
 (Hint: For (a), first show that $A_2(6, 5) \geq 2$ by producing a code explicitly.

Then try to show that $A_2(6, 5) \leq 2$ using a simple combinatorial argument similar to the one in Example 5.2.5.)

5.5 Find an optimal binary code with $n = 3$ and $d = 2$.

5.6 Prove that $A_q(n, d) \leq q A_q(n - 1, d)$.

5.7 For each of the following spheres in $A^n = F_2^n$, list its elements and compute its volume:

(a) $S_A(110, 4)$,   (b) $S_A(1100, 3)$,   (c) $S_A(10101, 2)$.

5.8 For each n such that $4 \leq n \leq 12$, compute the Hamming bound and the sphere-covering bound for $A_2(n, 4)$.

5.9 Prove that a $(6, 20, 4)$-code over F_7 cannot be an optimal code.

5.10 Let $q \geq 2$ and $n \geq 2$ be any integers. Show that $A_q(n, 2) = q^{n-1}$.

5.11 Let C be an $[n, k, d]$-code over F_q, where $\gcd(d, q) = 1$. Suppose that all the codewords of C have weight congruent to 0 or d modulo q. Using Exercise 4.30(iv), or otherwise, show the existence of an $[n+1, k, d+1]$-code over F_q.

5.12 Let C be an optimal code over F_{11} of length 12 and minimum distance 2. Show that C must have a transmission rate of at least $5/6$.

5.13 For positive integers n, M, d and $q > 1$ (with $1 \leq d \leq n$), show that, if $(M - 1)\sum_{i=0}^{d-1} \binom{n}{i}(q - 1)^i < q^n$, then there exists a q-ary (n, M)-code of minimum distance at least d. (Note: this is often known as the *Gilbert–Varshamov bound* for nonlinear codes.)

5.14 Determine whether each of the following codes exists. Justify your answer.

(a) A binary code with parameters $(8, 29, 3)$.

(b) A binary linear code with parameters $(8, 8, 5)$.

(c) A binary linear code with parameters $(8, 5, 5)$.

(d) A binary linear code with parameters $(24, 2^{12}, 8)$.

(e) A perfect binary linear code with parameters $(63, 2^{57}, 3)$.

5.15 Write down a parity-check matrix H for a binary Hamming code of length 15, where the jth column of H is the binary representation of j. Then use H to construct a syndrome look-up table and use it to decode the following words:

(a) 01010 01010 01000,

(b) 11100 01110 00111,

(c) 11001 11001 11000.

5.16 (i) Show that there exist no binary linear codes with parameters $[2^m, 2^m - m, 3]$, for any $m \geq 2$.

(ii) Let C be a binary linear code with parameters $[2^m, k, 4]$, for some $m \geq 2$. Show that $k \leq 2^m - m - 1$.

5.17 Prove Proposition 5.3.15.

5.18 (i) Let $n \geq 3$ be an integer. Show that there is an $[n, k, 3]$-code defined
           over $\mathbf{F}_q$ if and only if $q^{n-k} - 1 \geq (q-1)n$.
      (ii) Find the smallest n for which there exists a ternary $[n, 5, 3]$-code.

5.19 (i) Let $\mathbf{v}$ be a nonzero vector in $\mathbf{F}_q^r$. Show that the set of vectors in
           $\mathbf{F}_q^r$ orthogonal to $\mathbf{v}$, i.e., $\{\mathbf{v}\}^\perp$, forms a subspace of $\mathbf{F}_q^r$ of dimension
           $r - 1$.
      (ii) Let G be a generator matrix for the simplex code $S(r, q)$. Show that,
           for a given nonzero vector $\mathbf{v} \in \mathbf{F}_q^r$, there are exactly $(q^{r-1}-1)/(q-1)$
           columns $\mathbf{c}$ of G such that $\mathbf{v} \cdot \mathbf{c} = 0$.
      (iii) Using the observation that $S(r, q) = \{\mathbf{v}G : \mathbf{v} \in \mathbf{F}_q^r\}$, or otherwise,
           show that every nonzero codeword of $S(r, q)$ has weight q^{r-1}.
           (Hint: Use (ii) to determine the number of coordinates of $\mathbf{v}G$ that
           are equal to 0.)

5.20 Determine the Hamming weight enumerators of $\mathrm{Ham}(3, 2)$ and $S(3, 2)$.
     Verify that they satisfy the MacWilliams identity (see Exercise 4.49).

5.21 The ternary Hamming code $\mathrm{Ham}(2, 3)$ is also known as the *tetracode*.
      (i) Show that the tetracode is a self-dual MDS code.
      (ii) Without writing down all the elements of $\mathrm{Ham}(2, 3)$, determine the
           weights of all its codewords.
      (iii) Determine the Hamming weight enumerator of $\mathrm{Ham}(2, 3)$ and show
           that the MacWilliams identity (see Exercise 4.49) holds for $C = C^\perp = \mathrm{Ham}(2, 3)$.

5.22 Let $\mathcal{G}_6$ denote the hexacode defined in Exercise 4.10(b).
      (i) Show that $\mathcal{G}_6$ is a $[6, 3, 4]$-code over $\mathbf{F}_4$. (Hence, $\mathcal{G}_6$ is an MDS
           quaternary code.)
      (ii) Let $\mathcal{G}_6'$ be the code obtained from $\mathcal{G}_6$ by deleting the last coordinate
           from every codeword. Show that $\mathcal{G}_6'$ is a Hamming code over $\mathbf{F}_4$.

5.23 (i) Show that the all-one vector $(1, 1, \ldots, 1)$ is in the extended binary
           Golay code G_{24}.
      (ii) Deduce from (i) that G_{24} does not have any word of weight 20.

5.24 Prove Proposition 5.3.22.

5.25 (i) Show that every word of weight 4 in $\mathbf{F}_2^{23}$ is of distance 3 from exactly
           one codeword in the binary Golay code G_{23}.
      (ii) Use (i) to count the number of codewords of weight 7 in G_{23}.
      (iii) Use (ii) to show that the extended binary Golay code G_{24} contains
           precisely 759 codewords of weight 8.

5.26 Show that the extended binary Golay code G_{24} has the weight distribution
     shown in Table 5.5 for its codewords.

5.27 Verify the MacWilliams identity (see Exercise 4.49) with $C = C^\perp = G_{24}$.

Table 5.5.

Weight	0	4	8	12	16	20	24
Number of codewords	1	0	759	2576	759	0	1

5.28 Prove that the extended ternary Golay code G_{12} is a $[12, 6, 6]$-code.

5.29 Show that the ternary Golay code G_{11} satisfies the Hamming bound.

5.30 Prove Theorem 5.5.3. (Hint: When d is even and $n < 2d$, mimic the proof of Theorem 5.5.2. Divide into the two cases M even and M odd, and maximize the expression $\sum_{i=1}^{n} \sum_{a \in \mathbf{F}_2} n_{i,a}(M - n_{i,a})$ in each case. For the case of even d and $n = 2d$, apply Exercise 5.6, with $q = 2$, and the previous case. When d is odd, apply Theorem 5.1.11 with the result for even d.)

5.31 Let C be the code over $\mathbf{F}_4 = \{0, 1, \alpha, \alpha^2\}$ with generator matrix

$$\begin{pmatrix} 1 & 0 & 1 & 1 \\ 0 & 1 & \alpha & \alpha^2 \end{pmatrix}.$$

(i) Show that C is an MDS code.

(ii) Write down a generator matrix for the dual $C^\perp$.

(iii) Show that $C^\perp$ is an MDS code.

5.32 Show that the only binary MDS codes are the trivial ones.

5.33 Suppose there is a q-ary MDS code C of length n and dimension k, where $k < n$.

(i) Show that there is also a q-ary MDS code of length $n - 1$ and dimension k.

(ii) For a given $1 \leq i \leq n$, let C_i be the subcode of C consisting of all the codewords with 0 in the ith position, and let D_i be the code obtained by deleting the ith coordinate from every codeword of C_i. Show that D_i is an MDS code. (Hint: You may need to show that there is at least one minimum weight codeword of C with 0 in the ith position.)

5.34 For each n such that $9 \leq n \leq 16$, compare the Singleton, Plotkin and Hamming upper bounds for $A_2(n, 9)$.

5.35 Suppose there exists a binary linear code C of parameters $[16, 8, 6]$.

(i) Let C' be the residual code of C with respect to a codeword of weight 6. Show that C' is a binary linear code of parameters $[10, 7, d']$, where $3 \leq d' \leq 4$.

(ii) Use Exercise 5.32 to show that $d' = 3$.

(iii) Using the Hamming bound, or otherwise, show that such a C' cannot exist.

5.36 A binary (n, M, d)-code C is called a *constant-weight binary code* if there exists an integer w such that wt($\mathbf{c}$) $= w$ for all $\mathbf{c} \in C$. In this case, we say that C is a constant-weight binary $(n, M, d; w)$-code.

(a) Show that the minimum distance of a constant-weight binary code is always even.

(b) Show that a constant-weight binary $(n, M, d; w)$-code satisfies $M \le \binom{n}{w}$.

(c) Prove that a constant-weight binary $(n, M, d; w)$-code can detect at least one error.

5.37 Let $A_2(n, d, w)$ be the maximum possible number M of codewords in a constant-weight binary $(n, M, d; w)$-code. Show that

(a) $1 \le A_2(n, d, w) \le \binom{n}{w}$;

(b) $A_2(n, 2, w) = \binom{n}{w}$;

(c) $A_2(n, d, w) = 1$ for $d > 2w$;

(d) $A_2(n, d, w) = A_2(n, d, n - w)$.

5.38 Use the Griesmer bound to find an upper bound for d for the q-ary linear codes of the following n and k:

(a) $q = 2$, $n = 10$ and $k = 3$;

(b) $q = 3$, $n = 8$ and $k = 4$;

(c) $q = 4$, $n = 10$ and $k = 5$;

(d) $q = 5$, $n = 9$ and $k = 2$.

5.39 For a prime power q and positive integers k and u with $k > u > 0$, the *MacDonald code* $C_{k,u}$ is a q-ary linear code, of parameters $[(q^k - q^u)/(q - 1), k, q^{k-1} - q^{u-1}]$, that has nonzero codewords of only two possible weights: $q^{k-1} - q^{u-1}$ and q^{k-1}. Show that the MacDonald codes attain the Griesmer bound.

5.40 Let C be an $[n, k, d]$-code over $\mathbf{F}_q$ and let $\mathbf{c} \in C$ be a codeword of weight w, where $w < dq/(q - 1)$. Show that the residual code Res(C, $\mathbf{c}$) is an $[n - w, k - 1, d']$-code, where $d' \ge d - w + \lceil w/q \rceil$.

5.41 Let C be a $[q^2, 4, q^2 - q - 1]$-code over $\mathbf{F}_q$.

(i) By considering Res(C, $\mathbf{c}$), where wt($\mathbf{c}$) $= q^2 - t$ with $2 \le t \le q - 1$, or otherwise, show that the only possible weights of the codewords in C are: 0, $q^2 - q - 1$, $q^2 - q$, $q^2 - 1$ and q^2.

(ii) Show the existence of a $[q^2 + 1, 4, q^2 - q]$-code over $\mathbf{F}_q$. (Hint: Compare with Exercise 5.11.)

5.42 Prove the properties of the Krawtchouk polynomials listed in Proposition 5.8.2. (Hint: For (ii), use the fact that

$$(1 + (q-1)z)^{n-x}(1-z)^x = (1-z)^n \left(1 + \frac{qz}{1-z}\right)^{n-x}.$$

For (iv), multiply both sides of the equality by $y^k z^\ell$ and sum over all $k, \ell \geq 0$. For (vii), use (ii). For (viii), use (i) by multiplying two power series together.)

5.43 Show that the Krawtchouk polynomials satisfy the following recurrence relation:

$$(k+1)K_{k+1}(x)$$
$$= (k + (q-1)(n-k) - qx)\,K_k(x) - (q-1)(n-k+1)K_{k-1}(x).$$

5.44 Show that $K_k(x) = \sum_{j=0}^{k}(-q)^j(q-1)^{k-j}\binom{n-j}{k-j}\binom{x}{j}$.

5.45 Let $q = 2$. Show that:
   (a) $K_0(x) = 1$;
   (b) $K_1(x) = -2x + n$;
   (c) $K_2(x) = 2x^2 - 2nx + \binom{n}{2}$;
   (d) $K_3(x) = -4x^3/3 + 2nx^2 - (n^2 - n + 2/3)x + \binom{n}{3}$.

5.46 Let ζ be a primitive qth root of unity in $\mathbf{C}$. Suppose $\mathbf{u} \in \mathbf{Z}_q^n$ is a word of weight i. Show that

$$\sum_{\substack{\mathbf{w} \in \mathbf{Z}_q^n \\ \mathrm{wt}(\mathbf{w})=k}} \zeta^{\mathbf{u}\cdot\mathbf{w}} = K_k(i),$$

where, for $\mathbf{u} = (u_1, \ldots, u_n)$ and $\mathbf{w} = (w_1, \ldots, w_n)$, $\mathbf{u}\cdot\mathbf{w} = u_1 w_1 + \cdots + u_n w_n$.

5.47 Use the linear programming bound (Theorem 5.8.7) to show that the Hadamard matrix code of parameters $(2d, 4d, d)$, with d even, is an optimal code. (Hint: Use $f(x) = 1 + K_1(x) + \frac{1}{d}K_2(x)$.)

5.48 Let d be such that $2d > n$. Use the linear programming bound (Theorem 5.8.7) to show that $A_2(n, d) \leq 2d/(2d - n)$. Note that this bound is slightly weaker than the Plotkin bound. (Hint: Use $f(x) = 1 + \frac{1}{2d-n}K_1(x)$.)

6 Constructions of linear codes

For an (n, M, d)-code C over $\mathbf{F}_q$, theoretically we would like both $\mathcal{R}(C) = (\log_q M)/n$ and $\delta(C) = (d-1)/n$ to be as large as possible. In other words, we want M to be as large as possible for fixed n and d. Of course, the ideal situation is to find codes with size equal to $A_q(n, d)$ for all given q, n and d. However, from the previous chapter, we know that it is still an open problem that seems difficult to solve. Fortunately, in practice, we are contented to use codes with sizes close to $A_q(n, d)$. In order to do so, we have to find ways to construct such codes.

The construction of good codes has almost as long a history as coding theory itself. The Hamming codes in the previous chapter are one of the earliest classes of codes. Later on, many other codes which are also in practical use were invented: for instance, Reed–Muller codes (see Section 6.2); BCH codes (Section 8.1); Reed–Solomon codes (Section 8.2); and Goppa codes (Section 9.3).

In this chapter, we concentrate mainly on the construction of linear codes. For the construction of nonlinear codes, the reader is advised to refer to ref. [13] and the following webpage maintained by Simon Litsyn of Tel Aviv University, E. M. Rains of IDA and N. J. A. Sloane of AT&T Labs-Research:

http://www.research.att.com/~njas/codes/And/.

6.1 Propagation rules

In this section, we study several constructions of new codes based on old codes. Our strategy is to build codes with larger sizes or longer lengths from codes of smaller sizes or shorter lengths. From the beginning of coding theory, many propagation rules have been proposed, and some of them have become standard constructions in coding theory. We feature a few well known propagation rules in this section and place some others in the exercises to this chapter.

Theorem 6.1.1 *Suppose there is an $[n, k, d]$-linear code over $\mathbf{F}_q$. Then*

(i) (lengthening) *there exists an $[n + r, k, d]$-linear code over $\mathbf{F}_q$ for any $r \geq 1$;*

(ii) (subcodes) *there exists an $[n, k - r, d]$-linear code over $\mathbf{F}_q$ for any $1 \leq r \leq k - 1$;*

(iii) (puncturing) *there exists an $[n - r, k, d - r]$-linear code over $\mathbf{F}_q$ for any $1 \leq r \leq d - 1$;*

(iv) *there exists an $[n, k, d - r]$-linear code over $\mathbf{F}_q$ for any $1 \leq r \leq d - 1$;*

(v) *there exists an $[n - r, k - r, d]$-linear code over $\mathbf{F}_q$ for any $1 \leq r \leq k - 1$.*

Proof. Let C be an $[n, k, d]$-linear code over $\mathbf{F}_q$.

(i) By mathematical induction, it suffices to show the existence of an $[n+1, k, d]$-linear code over $\mathbf{F}_q$. We add a new coordinate 0 to all the codewords of C to form a new code,

$$\{(u_1, \ldots, u_n, 0) : (u_1, \ldots, u_n) \in C\}.$$

It is clear that the above code is an $[n + 1, k, d]$-linear code over $\mathbf{F}_q$.

(ii) Let $\mathbf{c}$ be a nonzero codeword of C with $\mathrm{wt}(\mathbf{c}) = d$. We extend $\mathbf{c}$ to form a basis of C: $\{\mathbf{c}_1 = \mathbf{c}, \ldots, \mathbf{c}_k\}$. Consider the new code $< \{\mathbf{c}_1, \ldots, \mathbf{c}_{k-r}\} >$ spanned by the first $k - r$ codewords in the basis. It is obvious that the new code has the parameters $[n, k - r, d]$.

(iii) Let $\mathbf{c} \in C$ be a codeword of weight d. For each codeword of C, we delete a fixed set of r positions where $\mathbf{c}$ has nonzero coordinates (compare with the proof of Theorem 5.1.11). It is easy to see that the new code is an $[n - r, k, d - r]$-linear code.

(iv) The desired result follows from (i) and (iii).

(v) If $k = n$, then we must have that $d = 1$. Thus, the space $\mathbf{F}_q^{n-r}$ is a code with parameters $[n - r, k - r, d]$.

Now we assume that $k < n$. It also suffices to show the existence of an $[n - 1, k - 1, d]$-linear code for $k \geq 2$. Let C be an $[n, k, d]$-linear code over $\mathbf{F}_q$. We may assume that C has a parity-check matrix of the form

$$H = (I_{n-k}|X).$$

Deleting the last column of H, we obtain an $(n - k) \times (n - 1)$ matrix H_1. It is clear that all the rows of H_1 are linearly independent and that any $d - 1$ columns of H_1 are linearly independent. Thus, the linear code with H_1 as a parity-check matrix has parameters $[n-1, k-1, d_1]$ with $d_1 \geq d$. By part (iv), we have an $[n-1, k-1, d]$-linear code. □

Remark 6.1.2 In fact, the above theorem produces codes with worse parameters than the old ones. We usually do not make new codes using these constructions. However, they are quite useful when we study codes.

The following result follows immediately from Theorem 6.1.1 (i)–(iv).

Corollary 6.1.3 *If there is an $[n, k, d]$-linear code over $\mathbf{F}_q$, then for any $r \geq 0$, $0 \leq s \leq k-1$ and $0 \leq t \leq d-1$, there exists an $[n+r, k-s, d-t]$-linear code over $\mathbf{F}_q$.*

Example 6.1.4 A binary Hamming code of length 7 is a $[7, 4, 3]$-linear code. Thus, we have binary linear codes with parameters $[n, 4, 3]$ for any $n \geq 7$ and also binary linear codes with parameters $[7, k, 3]$ for any $1 \leq k \leq 4$.

Theorem 6.1.5 (Direct sum.) *Let C_i be an $[n_i, k_i, d_i]$-linear code over $\mathbf{F}_q$ for $i = 1, 2$. Then the direct sum of C_1 and C_2 defined by*

$$C_1 \oplus C_2 = \{(\mathbf{c}_1, \mathbf{c}_2) : \mathbf{c}_1 \in C_1, \mathbf{c}_2 \in C_2\}$$

is an $[n_1 + n_2, k_1 + k_2, \min\{d_1, d_2\}]$-linear code over $\mathbf{F}_q$.

Proof. It is easy to verify that $C_1 \oplus C_2$ is a linear code over $\mathbf{F}_q$. The length of $C_1 \oplus C_2$ is clear. As the size of $C_1 \oplus C_2$ is equal to the product of the size of C_1 and that of C_2, we obtain

$$k := \dim(C_1 \oplus C_2) = \log_q(|C_1 \oplus C_2|) = \log_q(|C_1| \cdot |C_2|) = k_1 + k_2.$$

We may assume that $d_1 \leq d_2$. Let $\mathbf{u} \in C_1$ with $\mathrm{wt}(\mathbf{u}) = d_1$. Then $(\mathbf{u}, \mathbf{0}) \in C$. Hence, $d(C_1 \oplus C_2) \leq \mathrm{wt}((\mathbf{u}, \mathbf{0})) = d_1$. On the other hand, for any nonzero codeword $(\mathbf{c}_1, \mathbf{c}_2) \in C_1 \oplus C_2$ with $\mathbf{c}_1 \in C_1$ and $\mathbf{c}_2 \in C_2$, we have either $\mathbf{c}_1 \neq \mathbf{0}$ or $\mathbf{c}_2 \neq \mathbf{0}$. Thus,

$$\mathrm{wt}((\mathbf{c}_1, \mathbf{c}_2)) = \mathrm{wt}(\mathbf{c}_1) + \mathrm{wt}(\mathbf{c}_2) \geq d_1.$$

This completes the proof. □

Remark 6.1.6 Let G_i be a generator matrix of C_i, for $i = 1, 2$. Then it is easy to see that the matrix

$$\begin{pmatrix} G_1 & O \\ O & G_2 \end{pmatrix}$$

is a generator matrix of $C_1 \oplus C_2$, where O stands for the zero matrix (note that the two zero matrices have different sizes).

Example 6.1.7 Let

$$C_1 = \{000, 110, 101, 011\}$$

be a binary $[3, 2, 2]$-linear code, and let

$$C_2 = \{0000, 1111\}$$

be a binary $[4, 1, 4]$-linear code. Then,

$$C_1 \oplus C_2 = \{0000000, 1100000, 1010000, 0110000,$$
$$0001111, 1101111, 1011111, 0111111\}$$

is a binary $[7, 3, 2]$-linear code.

Theorem 6.1.8 (($\mathbf{u}, \mathbf{u} + \mathbf{v}$)-construction.) *Let C_i be an $[n, k_i, d_i]$-linear code over $\mathbf{F}_q$, for $i = 1, 2$. Then the code C defined by*

$$C = \{(\mathbf{u}, \mathbf{u} + \mathbf{v}) : \mathbf{u} \in C_1, \mathbf{v} \in C_2\}$$

is a $[2n, k_1 + k_2, \min\{2d_1, d_2\}]$-linear code over $\mathbf{F}_q$.

Proof. It is easy to verify that C is a linear code over $\mathbf{F}_q$. The length of C is clear.

It is easy to show that the map

$$C_1 \oplus C_2 \to C, \quad (\mathbf{c}_1, \mathbf{c}_2) \mapsto (\mathbf{c}_1, \mathbf{c}_1 + \mathbf{c}_2)$$

is a bijection. Thus, the size of C is equal to the product of the size of C_1 and that of C_2; i.e., $k := \dim C = k_1 + k_2$.

For any nonzero codeword $(\mathbf{c}_1, \mathbf{c}_1 + \mathbf{c}_2) \in C$ with $\mathbf{c}_1 \in C_1$ and $\mathbf{c}_2 \in C_2$, we have either $\mathbf{c}_1 \neq \mathbf{0}$ or $\mathbf{c}_2 \neq \mathbf{0}$.

Case (1) $\mathbf{c}_2 = \mathbf{0}$. In this case, we have $\mathbf{c}_1 \neq \mathbf{0}$. Thus,

$$\text{wt}((\mathbf{c}_1, \mathbf{c}_1 + \mathbf{c}_2)) = \text{wt}((\mathbf{c}_1, \mathbf{c}_1)) = 2\text{wt}(\mathbf{c}_1) \geq 2d_1 \geq \min\{2d_1, d_2\}.$$

Case (2) $\mathbf{c}_2 \neq \mathbf{0}$. Then

$$\text{wt}((\mathbf{c}_1, \mathbf{c}_1 + \mathbf{c}_2)) = \text{wt}(\mathbf{c}_1) + \text{wt}(\mathbf{c}_1 + \mathbf{c}_2)$$
$$\geq \text{wt}(\mathbf{c}_1) + (\text{wt}(\mathbf{c}_2) - \text{wt}(\mathbf{c}_1)) \quad \text{(see Lemma 4.3.6)}$$
$$= \text{wt}(\mathbf{c}_2) \geq d_2 \geq \min\{2d_1, d_2\}.$$

This shows that $d(C) \geq \min\{2d_1, d_2\}$. On the other hand, let $\mathbf{x} \in C_1$ and $\mathbf{y} \in C_2$ with $\text{wt}(\mathbf{x}) = d_1$ and $\text{wt}(\mathbf{y}) = d_2$. Then $(\mathbf{x}, \mathbf{x})$, $(\mathbf{0}, \mathbf{y}) \in C$ and

$$d(C) \leq \text{wt}((\mathbf{x}, \mathbf{x})) = 2d_1$$

and

$$d(C) \leq \text{wt}((\mathbf{0}, \mathbf{y})) = d_2.$$

Thus, $d(C) \leq \min\{2d_1, d_2\}$. This completes the proof. □

Remark 6.1.9 Let G_i be a generator matrix of C_i, for $i = 1, 2$. Then it is easy to see that the matrix

$$\begin{pmatrix} G_1 & G_1 \\ O & G_2 \end{pmatrix}$$

is a generator matrix of C from the $(\mathbf{u}, \mathbf{u} + \mathbf{v})$-construction in Theorem 6.1.8, where O stands for the zero matrix.

Example 6.1.10 Let

$$C_1 = \{000, 110, 101, 011\}$$

be a binary $[3, 2, 2]$-linear code, and let

$$C_2 = \{000, 111\}$$

be a binary $[3, 1, 3]$-linear code. Then,

$$C = \{000000, 110110, 101101, 011011,$$
$$000111, 110001, 101010, 011100\}$$

is a binary $[6, 3, 3]$-linear code.

Let $\mathbf{1} = (1, 1, \ldots, 1)$ denote the all-one vector and let $\mathbf{0} = (0, 0, \ldots, 0)$ denote the zero vector. (The length is unspecified here and depends on the context.)

Corollary 6.1.11 *Let A be a binary $[n, k, d]$-linear code. Then the code C defined by*

$$C = \{(\mathbf{c}, \mathbf{c}) : \mathbf{c} \in A\} \cup \{(\mathbf{c}, \mathbf{1} + \mathbf{c}) : \mathbf{c} \in A\}$$

is a binary $[2n, k + 1, \min\{n, 2d\}]$-linear code.

Proof. In Theorem 6.1.8, taking $C_1 = A$ and $C_2 = \{\mathbf{0}, \mathbf{1}\}$, we obtain the desired result. □

Example 6.1.12 Let

$$A = \{00, 01, 10, 11\}$$

be a binary $[2, 2, 1]$-linear code. Put

$$
\begin{aligned}
C &= \{(\mathbf{c}, \mathbf{c}) : \mathbf{c} \in A\} \cup \{(\mathbf{c}, \mathbf{1} + \mathbf{c}) : \mathbf{c} \in A\} \\
&= \{0000, 0101, 1010, 1111, 0011, 0110, 1001, 1100\}.
\end{aligned}
$$

Then C is a binary $[4, 3, 2]$-linear code.

6.2 Reed–Muller codes

Reed–Muller codes are among the oldest known codes and have found widespread applications. For each positive integer m and each integer r satisfying $0 \le r \le m$, there is an rth order Reed–Muller code $\mathcal{R}(r, m)$, which is a binary linear code of parameters $\left[2^m, \binom{m}{0} + \binom{m}{1} + \cdots + \binom{m}{r}, 2^{m-r}\right]$. In fact, $\mathcal{R}(1, 5)$ was used by Mariner 9 to transmit black and white pictures of Mars back to the Earth in 1972. Reed–Muller codes also admit a special decoding called the Reed decoding. There are also generalizations to nonbinary fields. We concentrate mainly on the first order binary Reed–Muller codes.

There are many ways to define the Reed–Muller codes. We choose an inductive definition. Remember that we are in the binary setting.

Definition 6.2.1 The (*first order*) *Reed–Muller codes* $\mathcal{R}(1, m)$ are binary codes defined, for all integers $m \ge 1$, recursively as follows:

(i) $\mathcal{R}(1, 1) = \mathbf{F}_2^2 = \{00, 01, 10, 11\}$;
(ii) for $m \ge 1$,

$$\mathcal{R}(1, m + 1) = \{(\mathbf{u}, \mathbf{u}) : \mathbf{u} \in \mathcal{R}(1, m)\} \cup \{(\mathbf{u}, \mathbf{u} + \mathbf{1}) : \mathbf{u} \in \mathcal{R}(1, m)\}.$$

Example 6.2.2 $\mathcal{R}(1, 2) = \{0000, 0101, 1010, 1111, 0011, 0110, 1001, 1100\}$. A generator matrix of $\mathcal{R}(1, 2)$ is

$$
\begin{pmatrix}
1 & 1 & 1 & 1 \\
0 & 1 & 0 & 1 \\
0 & 0 & 1 & 1
\end{pmatrix}.
$$

Proposition 6.2.3 *For* $m \ge 1$, *the Reed–Muller code* $\mathcal{R}(1, m)$ *is a binary* $[2^m, m + 1, 2^{m-1}]$-*linear code, in which every codeword except* $\mathbf{0}$ *and* $\mathbf{1}$ *has weight* 2^{m-1}.

Proof. It is clear that $\mathcal{R}(1, 1)$ is a binary $[2, 2, 1]$-linear code. We note that $\mathcal{R}(1, m)$ is obtained from $\mathcal{R}(1, m - 1)$ by the construction in Corollary 6.1.11. Using mathematical induction, we assume that $\mathcal{R}(1, m - 1)$ is a binary $[2^{m-1}, m, 2^{m-2}]$-linear code. By Corollary 6.1.11, $\mathcal{R}(1, m)$ is a binary $[2 \cdot 2^{m-1}, m + 1, \min\{2 \cdot 2^{m-2}, 2^{m-1}\}] = [2^m, m + 1, 2^{m-1}]$-linear code.

Now we prove that, except for $\mathbf{0}$ and $\mathbf{1}$, every codeword of $\mathcal{R}(1, m + 1)$ has weight $2^m = 2^{(m+1)-1}$.

A word in $\mathcal{R}(1, m + 1)$ is either of the type $(\mathbf{u}, \mathbf{u})$ or $(\mathbf{u}, \mathbf{u} + \mathbf{1})$, where $\mathbf{u}$ is a word in $\mathcal{R}(1, m)$.

Case (1) $(\mathbf{u}, \mathbf{u})$, where $\mathbf{u} \in \mathcal{R}(1, m)$: $\mathbf{u}$ can be neither $\mathbf{0}$ nor $\mathbf{1}$, since otherwise $(\mathbf{u}, \mathbf{u})$ is again the zero or the all-one vector. Hence, by the inductive hypothesis, $\mathbf{u}$ has weight 2^{m-1}. Therefore, $(\mathbf{u}, \mathbf{u})$ has weight $2 \cdot 2^{m-1} = 2^m$.

Case (2) $(\mathbf{u}, \mathbf{u} + \mathbf{1})$, where $\mathbf{u} \in \mathcal{R}(1, m)$:

(a) If $\mathbf{u}$ is neither $\mathbf{0}$ nor $\mathbf{1}$, then it has weight 2^{m-1}; i.e., exactly half its coordinates are 1. Hence, half of the coordinates of $\mathbf{u} + \mathbf{1}$ are 1; i.e., the weight of $\mathbf{u} + \mathbf{1}$ is also 2^{m-1}. Therefore, the weight of $(\mathbf{u}, \mathbf{u} + \mathbf{1})$ is exactly 2^m.
(b) If $\mathbf{u} = \mathbf{0}$, then $\mathbf{u} + \mathbf{1} = \mathbf{1}$, so again the weight of $(\mathbf{0}, \mathbf{0} + \mathbf{1})$ is 2^m.
(c) If $\mathbf{u} = \mathbf{1}$, then $\mathbf{u} + \mathbf{1} = \mathbf{0}$, so the weight of $(\mathbf{1}, \mathbf{1} + \mathbf{1})$ is again 2^m. □

Proposition 6.2.4 (i) *A generator matrix of* $\mathcal{R}(1, 1)$ *is*

$$\begin{pmatrix} 1 & 1 \\ 0 & 1 \end{pmatrix}.$$

(ii) *If* G_m *is a generator matrix for* $\mathcal{R}(1, m)$, *then a generator matrix for* $\mathcal{R}(1, m + 1)$ *is*

$$G_{m+1} = \begin{pmatrix} G_m & G_m \\ 0 \cdots 0 & 1 \cdots 1 \end{pmatrix}.$$

Proof. (i) is obvious, while (ii) is an immediate result of Corollary 6.1.11 and Remark 6.1.9. □

Example 6.2.5 Using the generator matrix

$$G_2 = \begin{pmatrix} 1 & 1 & 1 & 1 \\ 0 & 1 & 0 & 1 \\ 0 & 0 & 1 & 1 \end{pmatrix},$$

we have

$$G_3 = \begin{pmatrix} 1 & 1 & 1 & 1 & 1 & 1 & 1 & 1 \\ 0 & 1 & 0 & 1 & 0 & 1 & 0 & 1 \\ 0 & 0 & 1 & 1 & 0 & 0 & 1 & 1 \\ 0 & 0 & 0 & 0 & 1 & 1 & 1 & 1 \end{pmatrix}.$$

Proposition 6.2.6 *The dual code $\mathcal{R}(1, m)^\perp$ is (equivalent to) the extended binary Hamming code $\overline{\mathrm{Ham}(m, 2)}$.*

Proof. From Proposition 6.2.4(ii), starting with

$$G_1 = \begin{pmatrix} 1 & 1 \\ 0 & 1 \end{pmatrix},$$

it is clear that G_m is of the form

$$\left(\begin{array}{c|c} 1 & 1 \cdots 1 \\ \hline 0 & \\ \vdots & H_m \\ 0 & \end{array} \right),$$

where H_m is some matrix. Moving the first coordinate to the last and moving the first row of the matrix to the last, we obtain the following generator matrix G'_m for an equivalent code:

$$\left(\begin{array}{c|c} & 0 \\ H_m & \vdots \\ & 0 \\ \hline 1 \cdots 1 & 1 \end{array} \right).$$

Using Theorem 5.1.9, if we show that H_m is a parity-check matrix for $\mathrm{Ham}(m, 2)$, then G'_m is the parity-check matrix for $\overline{\mathrm{Ham}(m, 2)}$, so $\mathcal{R}(1, m)^\perp$ is equivalent to $\overline{\mathrm{Ham}(m, 2)}$.

To show H_m is a parity-check matrix for $\mathrm{Ham}(m, 2)$, we need to show that the columns of H_m consist of all the nonzero vectors of length m.

Indeed, when $m = 1, 2$, the columns of H_m consist of all the nonzero vectors of length m. Now suppose that the columns of H_m consist of all the nonzero vectors of length m, for some m. By the definition of G_m, it follows readily that the columns of H_{m+1} consist of the following:

$$\begin{pmatrix} \mathbf{c} \\ 0 \end{pmatrix}, \begin{pmatrix} \mathbf{c} \\ 1 \end{pmatrix}, \text{ and } \begin{pmatrix} \mathbf{0}^\mathrm{T} \\ 1 \end{pmatrix},$$

where **c** is one of the columns of H_m and **0** is the zero vector of length m. It is clear that the vectors in this list make up exactly all the nonzero vectors of length $m+1$. Hence, by induction, the columns of H_m consist of all the nonzero vectors of length m.   □

Finally, we define the rth order Reed–Muller codes.

Definition 6.2.7 (i) The *zeroth order Reed–Muller codes* $\mathcal{R}(0, m)$, for $m \geq 0$, are defined to be the repetition codes $\{\mathbf{0}, \mathbf{1}\}$ of length 2^m.
(ii) The *first order Reed–Muller codes* $\mathcal{R}(1, m)$, for $m \geq 1$, are defined as in Definition 6.2.1.
(iii) For any $r \geq 2$, the *rth order Reed–Muller codes* $\mathcal{R}(r, m)$ are defined, for $m \geq r - 1$, recursively by

$$\mathcal{R}(r, m+1) = \begin{cases} \mathbf{F}_2^{2^r} & \text{if } m = r - 1 \\ \{(\mathbf{u}, \mathbf{u} + \mathbf{v}) \ : \ \mathbf{u} \in \mathcal{R}(r, m), \\ \qquad \mathbf{v} \in \mathcal{R}(r-1, m)\} & \text{if } m > r - 1. \end{cases}$$

6.3   Subfield codes

For the two previous sections, we made use of codes over $\mathbf{F}_q$ to construct new ones over the same ground field. In this section, we will employ codes over an extension $\mathbf{F}_{q^m}$ to obtain codes over $\mathbf{F}_q$.

Theorem 6.3.1 (Concatenated code.) *Let A be an $[N, K, D]$-linear code over $\mathbf{F}_{q^m}$. Then there exists an $[nN, mK, d']$-linear code C over $\mathbf{F}_q$ with $d' = d(C) \geq dD$, provided that there is an $[n, m, d]$-linear code B over $\mathbf{F}_q$. Moreover, an $[nN, mK, dD]$-linear code over $\mathbf{F}_q$ can be obtained.*

Proof. As $\mathbf{F}_{q^m}$ can be viewed as an $\mathbf{F}_q$-vector space of dimension m, we set up an $\mathbf{F}_q$-linear transformation ϕ between $\mathbf{F}_{q^m}$ and B such that ϕ is bijective.
We extend the map ϕ and obtain a map

$$\phi^* : \quad \mathbf{F}_{q^m}^N \to \mathbf{F}_q^{nN}, \qquad (v_1, \ldots, v_N) \mapsto (\phi(v_1), \ldots, \phi(v_N)).$$

It is easy to see that ϕ^* is an $\mathbf{F}_q$-linear transformation from $\mathbf{F}_{q^m}^N$ to $\mathbf{F}_q^{nN}$. The map ϕ^* is one-to-one (but not onto unless $n = m$).
The code A is an $\mathbf{F}_q$-subspace of $\mathbf{F}_{q^m}^N$. Let C be the image of A under ϕ^*; i.e., $C = \phi^*(A)$. Then C is a subspace of $\mathbf{F}_q^{nN}$ since ϕ^* is an $\mathbf{F}_q$-linear transformation.
The length of C is clearly nN. To determine the dimension of C, we recall a relationship between the size of a vector space V over $\mathbf{F}_r$ and its dimension

(see Theorem 4.1.15(i)):

$$\dim_{\mathbf{F}_r} V = \log_r |V| \quad \text{or} \quad |V| = r^{\dim_{\mathbf{F}_r} V}. \tag{6.1}$$

Thus, we have

$$
\begin{aligned}
\dim_{\mathbf{F}_q} C &= \log_q |C| \\
&= \log_q |A| \qquad (\text{as } \phi^* \text{ is one-to-one}) \\
&= \log_q \left((q^m)^{\dim_{\mathbf{F}_{q^m}} A} \right) \\
&= \log_q q^{mK} = mK.
\end{aligned}
$$

Finally, we look at the minimum distance of C. Let $(u_1, \ldots, u_N)$ be a nonzero codeword of A. If $u_i \neq 0$ for some $1 \leq i \leq N$, then $\phi(u_i)$ is a nonzero codeword of B. Hence, $\mathrm{wt}(\phi(u_i)) \geq d$. As $(u_1, \ldots, u_N)$ has at least D nonzero positions, the number of nonzero positions of $(\phi(u_1), \ldots, \phi(u_N))$ is at least dD.

By Theorem 6.1.1(iv), we obtain an $[nN, mK, dD]$-linear code over $\mathbf{F}_q$. This completes the proof. $\square$

The code A in Theorem 6.3.1 is called the *outer code*, while the code B in Theorem 6.3.1 is called the *inner code*.

In Theorem 6.3.1, let $B = \mathbf{F}_q^m$ be the trivial code with the parameters $[m, m, 1]$. We obtain the following result.

Corollary 6.3.2 *We have an $[mN, mK, D]$-linear code over $\mathbf{F}_q$ whenever there is an $[N, K, D]$-linear code over $\mathbf{F}_{q^m}$.*

Example 6.3.3 (i) We know that there exist a $[17, 15, 3]$-Hamming code over $\mathbf{F}_{16}$ and a binary $[8, 4, 4]$-linear code (see Proposition 5.3.10). By Theorem 6.3.1, we obtain a binary $[136, 60, 12]$-linear code.

(ii) We have an $[(8^3 - 1)/(8 - 1), (8^3 - 1)/(8 - 1) - 3, 3] = [73, 70, 3]$-Hamming code over $\mathbf{F}_8$. Thus, we obtain a binary $[219, 210, 3]$-linear code by Corollary 6.3.2.

Example 6.3.4 (i) Consider the linear code

$$A := \{(0, 0), (1, \alpha), (\alpha, 1 + \alpha), (1 + \alpha, 1)\}$$

over $\mathbf{F}_4$, where α is a root of $1 + x + x^2$. Let B be the binary code

$$\{000, 110, 101, 011\},$$

and consider the $\mathbf{F}_2$-linear transformation between $\mathbf{F}_4$ and B defined by

$$\phi: \quad 0 \mapsto 000, \quad 1 \mapsto 110, \quad \alpha \mapsto 101, \quad 1 + \alpha \mapsto 011.$$

Then we obtain the code

$$C := \phi^*(A) = \{000000, 110101, 101011, 011110\}.$$

The new code C has parameters $[6, 2, 4]$.

(ii) Let α be a root of $1 + x + x^3 \in F_2[x]$. Then, $F_8 = F_2[\alpha]$. By Exercise 4.4, $\{1, \alpha, \alpha^2\}$ forms a basis of F_8 over F_2. Consider the map $\phi : F_8 \to F_2^3$

$$a_1 \cdot 1 + a_2 \cdot \alpha + a_3 \cdot \alpha^2 \mapsto (a_1, a_2, a_3).$$

Let $A =< (\alpha, \alpha + 1, 1) > /F_8$. By Corollary 6.3.2, $C := \phi^*(A)$ is a binary $[9, 3, d]$-linear code, where $d \geq 3$. We list all the elements of A:

$$A = \{(0, 0, 0), (\alpha, \alpha + 1, 1), (\alpha^2, \alpha^2 + \alpha, \alpha),$$
$$(\alpha + 1, \alpha^2 + \alpha + 1, \alpha^2), (\alpha^2 + \alpha, \alpha^2 + 1, \alpha + 1),$$
$$(\alpha^2 + \alpha + 1, 1, \alpha^2 + \alpha), (\alpha^2 + 1, \alpha, \alpha^2 + \alpha + 1), (1, \alpha^2, \alpha^2 + 1)\}.$$

Therefore,

$$C = \phi^*(A) = \{000000000, 010110100, 001011010, 110111001,$$
$$011101110, 111100011, 101010111, 100001101\}.$$

Thus, C is in fact a binary $[9, 3, 4]$-linear code.

Any vector space V over F_{q^m} can be viewed as a vector space over F_q. In particular, $F_{q^m}^N$ is a vector space over F_q of dimension mN. This view brings out another construction.

Theorem 6.3.5 (Subfield subcode.) *Let C be an $[N, K, D]$-linear code over F_{q^m}. Then the subfield subcode $C|_{F_q} := C \cap F_q^N$ is an $[n, k, d]$-linear code over F_q with $n = N$, $k \geq mK - (m - 1)N$ and $d \geq D$. Moreover, an $[N, mK - (m - 1)N, D]$-linear code over F_q can be obtained provided that $mK > (m - 1)N$.*

Proof. It is clear that $C|_{F_q}$ is a linear code over F_q as both C and F_q^N can be viewed as F_q-subspaces of $F_{q^m}^N$.

The length of $C|_{F_q}$ is clear. For the dimension, we have

$$\dim_{F_q} C|_{F_q} = \dim_{F_q} (C \cap F_q^N)$$
$$= \dim_{F_q} C + \dim_{F_q} F_q^N - \dim_{F_q} (C + F_q^N)$$
$$\geq \log_q |C| + N - \dim_{F_q} (F_{q^m}^N)$$
$$\text{(as } C + F_q^N \text{ is an } F_q\text{-subspace of } F_{q^m}^N)$$
$$= \log_q (q^m)^K + N - \log_q q^{mN}$$
$$= mK + N - mN = mK - (m - 1)N.$$

As $C|_{F_q}$ is a subset of C, it is clear that the minimum Hamming weight of $C|_{F_q}$ is at least the minimum Hamming weight of C; i.e., $d(C|_{F_q}) \geq d(C) = D$.

Applying Corollary 6.1.3 gives the desired result on the second part. $\quad\square$

Example 6.3.6 Let α be a root of $1 + x + x^2 \in F_2[x]$. Then $F_4 = F_2[\alpha]$. Let

$$C = <\{(\alpha, 0, 0), (0, \alpha + 1, 0)\}> /F_4.$$

Thus, by Theorem 6.3.5, $C|_{F_2}$ is a binary $[3, k, d]$-linear code with

$$k \geq mK - (m-1)N = 2 \cdot 2 - (2-1) \cdot 3 = 1, \quad d \geq d(C) = 1.$$

We list all the elements of C:

$$\begin{aligned}
C = \{&(0, 0, 0), (\alpha, 0, 0), (1, 0, 0), (\alpha + 1, 0, 0) \\
&(0, \alpha + 1, 0), (0, \alpha, 0), (0, 1, 0), (\alpha, \alpha + 1, 0) \\
&(\alpha, \alpha, 0), (\alpha, 1, 0), (1, \alpha + 1, 0), (1, \alpha, 0) \\
&(1, 1, 0), (\alpha + 1, \alpha + 1, 0), (\alpha + 1, \alpha, 0) \\
&(\alpha + 1, 1, 0)\}.
\end{aligned}$$

It is clear that $C|_{F_2} = C \cap F_2^3 = \{000, 100, 010, 110\}$. Hence, $C|_{F_2}$ is in fact a binary $[3, 2, 1]$-linear code.

For the final construction in this section, we need the results of Exercise 4.5.

Theorem 6.3.7 (Trace code.) *Let C be an $[N, K, D]$-linear code over F_{q^m}. Then the trace code of C defined by*

$$\mathrm{Tr}_{F_{q^m}/F_q}(C) := \{(\mathrm{Tr}_{F_{q^m}/F_q}(c_1), \ldots, \mathrm{Tr}_{F_{q^m}/F_q}(c_n)) : (c_1, \ldots, c_n) \in C\}$$

is an $[n, k]$-linear code over F_q with $n = N$ and $k \leq mK$.

Proof. Since $\mathrm{Tr}_{F_{q^m}/F_q}$ is an F_q-linear transformation from F_{q^m} to F_q, the set $\mathrm{Tr}_{F_{q^m}/F_q}(C)$ is a subspace of F_q^n. It is clear that the length of $\mathrm{Tr}_{F_{q^m}/F_q}(C)$ is n. For the dimension, we have

$$\begin{aligned}
\dim_{F_q} \mathrm{Tr}_{F_{q^m}/F_q}(C) &= \log_q |\mathrm{Tr}_{F_{q^m}/F_q}(C)| \\
&\leq \log_q |C| = \log_q(q^m)^{\dim_{F_{q^m}} C} \\
&= \log_q(q^{mK}) = mK.
\end{aligned}$$

$\quad\square$

Example 6.3.8 Consider the code $C = \{\lambda(1, \alpha, \alpha + 1) : \lambda \in F_9\}$ over F_9, where α is a root of $2 + x + x^2 \in F_3[x]$. Then

$$C = \{(0, 0, 0), (\alpha, 1 + 2\alpha, 1), (1 + 2\alpha, 2 + 2\alpha, \alpha),$$
$$(2 + 2\alpha, 2, 1 + 2\alpha), (2, 2\alpha, 2 + 2\alpha), (2\alpha, 2 + \alpha, 2),$$
$$(2 + \alpha, 1 + \alpha, 2\alpha), (1 + \alpha, 1, 2 + \alpha), (1, \alpha, 1 + \alpha)\}.$$

Under the trace map Tr_{F_9/F_3}, we have

$$(0, 0, 0) \mapsto (0, 0, 0), \qquad\qquad (\alpha, 1 + 2\alpha, 1) \mapsto (2, 0, 2),$$
$$(1 + 2\alpha, 2 + 2\alpha, \alpha) \mapsto (0, 2, 2), \quad (2 + 2\alpha, 2, 1 + 2\alpha) \mapsto (2, 1, 0),$$
$$(2, 2\alpha, 2 + 2\alpha) \mapsto (1, 1, 2), \qquad (2\alpha, 2 + \alpha, 2) \mapsto (1, 0, 1),$$
$$(2 + \alpha, 1 + \alpha, 2\alpha) \mapsto (0, 1, 1), \qquad (1 + \alpha, 1, 2 + \alpha) \mapsto (1, 2, 0),$$
$$(1, \alpha, 1 + \alpha) \mapsto (2, 2, 1).$$

Hence, the trace code

$$\mathrm{Tr}_{F_9/F_3}(C) = \{(0, 0, 0), (2, 0, 2), (0, 2, 2), (2, 1, 0), (1, 1, 2),$$
$$(1, 0, 1), (0, 1, 1), (1, 2, 0), (2, 2, 1)\}$$

is a $[3, 2, 2]$-linear code over $\mathbf{F}_3$.

In fact, trace codes are none other than subfield subcodes. This is shown by the following result.

Theorem 6.3.9 (Delsarte.) *For a linear code C over $\mathbf{F}_{q^m}$, one has*

$$(C|_{\mathbf{F}_q})^{\perp} = \mathrm{Tr}_{\mathbf{F}_{q^m}/\mathbf{F}_q}(C^{\perp}).$$

Proof. In order to prove that $(C|_{\mathbf{F}_q})^{\perp} \supseteq \mathrm{Tr}_{\mathbf{F}_{q^m}/\mathbf{F}_q}(C^{\perp})$, we have to show that

$$\mathbf{c} \cdot \mathrm{Tr}_{\mathbf{F}_{q^m}/\mathbf{F}_q}(\mathbf{a}) = 0 \quad \text{for all } \mathbf{a} \in C^{\perp} \text{ and } \mathbf{c} \in C|_{\mathbf{F}_q}.$$

Write $\mathbf{c} = (c_1, \ldots, c_n)$ and $\mathbf{a} = (a_1, \ldots, a_n)$; then

$$\mathbf{c} \cdot \mathrm{Tr}_{\mathbf{F}_{q^m}/\mathbf{F}_q}(\mathbf{a}) = \sum_{i=1}^{n} c_i \mathrm{Tr}_{\mathbf{F}_{q^m}/\mathbf{F}_q}(a_i) = \mathrm{Tr}_{\mathbf{F}_{q^m}/\mathbf{F}_q}\left(\sum_{i=1}^{n} c_i a_i\right)$$
$$= \mathrm{Tr}_{\mathbf{F}_{q^m}/\mathbf{F}_q}(\mathbf{c} \cdot \mathbf{a}) = 0.$$

We have used here the $\mathbf{F}_q$-linearity of the trace and the fact that $\mathbf{c} \cdot \mathbf{a} = 0$.

Next, we show that $(C|_{\mathbf{F}_q})^{\perp} \subseteq \mathrm{Tr}_{\mathbf{F}_{q^m}/\mathbf{F}_q}(C^{\perp})$. This assertion is equivalent to

$$\left(\mathrm{Tr}_{\mathbf{F}_{q^m}/\mathbf{F}_q}(C^{\perp})\right)^{\perp} \subseteq C|_{\mathbf{F}_q}.$$

Suppose the above relationship does not hold, then there exist some

$$\mathbf{u} \in \left(\mathrm{Tr}_{\mathbf{F}_{q^m}/\mathbf{F}_q}(C^{\perp})\right)^{\perp} \backslash C|_{\mathbf{F}_q}$$

and $\mathbf{v} \in C^{\perp}$ with $\mathbf{u} \cdot \mathbf{v} \neq 0$. As $\mathrm{Tr}_{\mathbf{F}_{q^m}/\mathbf{F}_q}$ is not the zero-map (see Exercise 4.5), there is an element $\gamma \in \mathbf{F}_{q^m}$ such that $\mathrm{Tr}_{\mathbf{F}_{q^m}/\mathbf{F}_q}(\gamma(\mathbf{u} \cdot \mathbf{v})) \neq 0$. Hence,

$$\mathbf{u} \cdot \mathrm{Tr}_{\mathbf{F}_{q^m}/\mathbf{F}_q}(\gamma \mathbf{v}) = \mathrm{Tr}_{\mathbf{F}_{q^m}/\mathbf{F}_q}(\mathbf{u} \cdot \gamma \mathbf{v}) = \mathrm{Tr}_{\mathbf{F}_{q^m}/\mathbf{F}_q}(\gamma(\mathbf{u} \cdot \mathbf{v})) \neq 0.$$

But, on the other hand, we have $\mathbf{u} \cdot \mathrm{Tr}_{\mathbf{F}_{q^m}/\mathbf{F}_q}(\gamma \mathbf{v}) = 0$ because $\mathbf{u} \in \left(\mathrm{Tr}_{\mathbf{F}_{q^m}/\mathbf{F}_q}(C^{\perp})\right)^{\perp}$ and $\gamma \mathbf{v} \in C^{\perp}$. The desired result follows from this contradiction. $\qquad\qquad\qquad\qquad\qquad\qquad\qquad\qquad\qquad\qquad\qquad\qquad\qquad\quad$ $\square$

The above theorem shows that trace codes can be obtained from subfield subcodes.

Example 6.3.10 As in Example 6.3.8, consider the code $C = \{\lambda(1, \alpha, \alpha + 1) : \lambda \in \mathbf{F}_9\}$ over $\mathbf{F}_9$, where α is a root of $2 + x + x^2 \in \mathbf{F}_3[x]$. Then, by Theorem 6.3.9 and Example 6.3.8, we have

$$
\begin{aligned}
C^{\perp}|_{\mathbf{F}_3} &= (\mathrm{Tr}_{\mathbf{F}_9/\mathbf{F}_3}(C))^{\perp} \\
&= \{(0, 0, 0), (2, 0, 2), (0, 2, 2), (2, 1, 0), (1, 1, 2), \\
&\qquad (1, 0, 1), (0, 1, 1), (1, 2, 0), (2, 2, 1)\}^{\perp} \\
&= \{(0, 0, 0), (1, 1, 2), (2, 2, 1)\}.
\end{aligned}
$$

Exercises

6.1 (a) Given an $[n, k, d]$-linear code over $\mathbf{F}_q$, can one always construct an $[n + 1, k + 1, d]$-linear code? Justify your answer.

(b) Given an $[n, k, d]$-linear code over $\mathbf{F}_q$, can one always construct an $[n + 1, k, d + 1]$-linear code? Justify your answer.

6.2 Let C be a q-ary $[n, k, d]$-linear code. For a fixed $1 \le i \le n$, form the subset A of C consisting of the codewords with the ith position equal to 0. Delete the ith position from all the words in A to form a code D. Show that D is a q-ary $[n - 1, k', d']$-linear code with

$$
k - 1 \le k' \le k, \quad d' \ge d.
$$

(Note: this way of obtaining a new code is called *shortening*.)

6.3 Suppose that

$$
G = \begin{pmatrix} 1 & 1 & 1 & 1 & 0 \\ 0 & 1 & 0 & 1 & 1 \\ 0 & 0 & 1 & 1 & 1 \end{pmatrix}
$$

is a generator matrix of a binary code C. Find a generator matrix of A with respect to $i = 2$ using the construction in Exercise 6.2.

6.4 Let H_i be a parity-check matrix of C_i, for $i = 1, 2$.

(a) Find a parity-check matrix of $C_1 \oplus C_2$ and justify your answer.

(b) Find a parity-check matrix of the code obtained from the $(\mathbf{u}, \mathbf{u} + \mathbf{v})$-construction and justify your answer.

6.5 (i) Let $A = \{0000, 1100, 0011, 1111\}$ be a binary code. Find the code C constructed from A using Corollary 6.1.11.

(ii) Let H be a parity-check matrix of A in (i). Find a parity-check matrix of C constructed from A using Corollary 6.1.11.

6.6 Assume that q is odd. Let C_i be an $[n, k_i, d_i]$-linear code over $\mathbf{F}_q$, for $i = 1, 2$. Define

$$C_1 \Diamond C_2 := \{(\mathbf{c}_1 + \mathbf{c}_2, \mathbf{c}_1 - \mathbf{c}_2) : \mathbf{c}_1 \in C_1, \mathbf{c}_2 \in C_2\}.$$

(a) Show that $C_1 \Diamond C_2$ is a $[2n, k_1 + k_2]$-linear code over $\mathbf{F}_q$.

(b) If G_i is a generator matrix of C_i, for $i = 1, 2$, find a generator matrix for $C_1 \Diamond C_2$ in terms of G_1 and G_2.

(c) Let d be the distance of $C_1 \Diamond C_2$. Show that $d = 2d_2$ if $2d_2 \leq d_1$ and $d_1 \leq d \leq 2d_2$ if $2d_2 > d_1$.

6.7 Let C_i be an $[n, k_i, d_i]$-linear code over $\mathbf{F}_q$, for $i = 1, 2$. Define

$$C := \{(\mathbf{a} + \mathbf{x}, \mathbf{b} + \mathbf{x}, \mathbf{a} + \mathbf{b} + \mathbf{x}) : \mathbf{a}, \mathbf{b} \in C_1, \mathbf{x} \in C_2\}.$$

(a) Show that C is a $[3n, 2k_1 + k_2]$-linear code over $\mathbf{F}_q$.

(b) If G_i is a generator matrix of C_i, for $i = 1, 2$, find a generator matrix of C in terms of G_1 and G_2.

(c) If H_i is a parity-check matrix of C_i, for $i = 1, 2$, find a parity-check matrix of C in terms of H_1 and H_2.

6.8 (a) Find the smallest n such that there exists a binary $[n, 50, 3]$-linear code.

(b) Find the smallest n such that there exists a binary $[n, 60, 4]$-linear code.

6.9 Find the smallest n such that there exists an $[n, 40, 3]$-linear code over $\mathbf{F}_9$.

6.10 (i) Write down the codewords in $\mathcal{R}(1, m)$ for $m = 3, 4, 5$.

(ii) Verify that $\mathcal{R}(1, 3)$ is self-dual.

6.11 Show that $\mathcal{R}(r, m)$ has parameters $\left[2^m, \binom{m}{0} + \binom{m}{1} + \cdots + \binom{m}{r}, 2^{m-r}\right]$.

6.12 For $0 \leq r < m$, show that $\mathcal{R}(r, m)^{\perp} = \mathcal{R}(m - 1 - r, m)$.

6.13 Write the binary solutions of the equation

$$x_1 + \cdots + x_m = 1$$

as column vectors of $\mathbf{F}_2^m$. Let $\mathbf{v}_1, \ldots, \mathbf{v}_n$ be all the solutions of the above equation. Let C_m be the binary linear code with

$$G = (\mathbf{v}_1, \ldots, \mathbf{v}_n)$$

as a generator matrix.

(i) Determine all the codewords of C_m, for $m = 2, 3, 4$.

(ii) Find the parameters of C_m for all m.

6.14 For a linear code V over $\mathbf{F}_q$, the parameters of V are denoted by

$$\text{length}(V), \dim(V) \text{ and } d(V) := \text{minimum distance}.$$

Suppose we have

(1) a code C with length $(C) = m$ and $\dim(C) = k$, and

(2) a collection of k codes $W_1, \ldots, W_k$, all of them having the same length n.

The elements of C are written as row vectors, and the elements of W_j are written as column vectors. We fix a basis $\{\mathbf{c}^{(1)}, \ldots, \mathbf{c}^{(k)}\}$ of C and denote by G the $k \times m$ matrix whose rows are $\mathbf{c}^{(1)}, \ldots, \mathbf{c}^{(k)}$. Thus, G is a generator matrix of C. For $1 \le j \le k$, we set

$$C_j := <\{\mathbf{c}^{(1)}, \ldots, \mathbf{c}^{(j)}\} > \subseteq \mathbf{F}_q^m.$$

Then C_j is a q-ary code of length m and dimension j. Moreover,

$$C_1 \subset C_2 \subset \cdots \subset C_k = C.$$

Let M be the set consisting of all the $n \times k$ matrices whose jth column belongs to W_j, for all $1 \le j \le k$.

(i) Show that M is an $\mathbf{F}_q$-linear space of dimension $\sum_{i=1}^{k} \dim(W_j)$.

(ii) If we identify an $n \times m$ matrix A with a vector $\mathbf{a}$ of $\mathbf{F}_q^{mn}$ by putting the ith row of A in the ith block of m positions of $\mathbf{a}$, then the q-ary linear code

$$W := \{AG : A \in M\}$$

has parameters

$$\begin{aligned}
\text{length}(W) &= mn, \\
\dim(W) &= \textstyle\sum_{j=1}^{k} \dim(W_j), \\
d(W) &\ge \min\{d(W_j) \cdot d(C_j) : 1 \le j \le k\}.
\end{aligned}$$

(iii) By using the binary codes with parameters $[2, 1, 2]$, $[20, 19, 2]$ and $[20, 14, 4]$, show that we can produce a binary $[40, 33, 4]$-linear code.

6.15 (i) Show that there always exists an $[n, n-1, 2]$-linear code over $\mathbf{F}_q$ for any $n \ge 2$.

(ii) Prove that there is an $[nN, (n-1)K, 2D]$-linear code over $\mathbf{F}_q$ whenever there is an $[N, K, D]$-linear code over $\mathbf{F}_{q^{n-1}}$.

6.16 Let α be a root of $1 + x^2 + x^3 \in \mathbf{F}_2[x]$. Consider the map

$$\phi : \mathbf{F}_8 \to \mathbf{F}_2^3, \quad a_1 \cdot 1 + a_2 \cdot \alpha + a_3 \cdot \alpha^2 \mapsto (a_1, a_2, a_3).$$

Let $A =< (\alpha + 1, \alpha^2 + 1, 1) > /\mathbf{F}_8$. Determine all the codewords of $\phi^*(A) = \{(\phi(c_1), \phi(c_2), \phi(c_3)) : (c_1, c_2, c_3) \in A\}$.

6.17 Consider the linear code

$$A :=< \{(1, 1), (\alpha, 1 + \alpha)\} >$$

over $\mathbf{F}_4$, where α is a root of $1 + x + x^2 \in \mathbf{F}_2[x]$. Let B be the binary code $\{0000, 1100, 1010, 0110\}$ and consider the $\mathbf{F}_2$-linear transformation between $\mathbf{F}_4$ and B defined by

$$\phi : \quad 0 \mapsto 0000, \quad 1 \mapsto 1100, \quad \alpha \mapsto 1010, \quad 1 + \alpha \mapsto 0110.$$

Determine all the codewords of the code

$$C := \phi^*(A) = \{(\phi(c_1), \phi(c_2)) : (c_1, c_2) \in A\}.$$

6.18 Let $\text{Ham}(m, 4)$ be a Hamming code of length $(4^m - 1)/3$ over $\mathbf{F}_4$. Using Theorem 6.3.5, estimate the parameters of $\text{Ham}(m, 4)|_{\mathbf{F}_2}$. Find the exact parameters of $\text{Ham}(3, 4)|_{\mathbf{F}_2}$.

6.19 Let $C =< \{(1, \alpha, \alpha^2), (\alpha^2, \alpha, 0)\} >$ be a linear code over $\mathbf{F}_4$, where α is a root of $1 + x + x^2 \in \mathbf{F}_2[x]$. Determine all the codewords of $C|_{\mathbf{F}_2}$.

6.20 (i) Suppose that $\mathbf{u}_1, \ldots, \mathbf{u}_r$ are vectors of $\mathbf{F}_q^n$. Show that the set $\{\mathbf{u}_1, \ldots, \mathbf{u}_r\}$ is $\mathbf{F}_q$-linearly independent if and only if it is $\mathbf{F}_{q^m}$-linearly independent for all $m \geq 1$.

   (ii) Show that, for an $[N, K]$-linear code C over $\mathbf{F}_{q^m}$, the subfield subcode $C|_{\mathbf{F}_q}$ has dimension at most K. Moreover, show that $\dim_{\mathbf{F}_q}(C|_{\mathbf{F}_q}) = K$ if and only if there is an $\mathbf{F}_{q^m}$-basis $\{\mathbf{c}_1, \ldots, \mathbf{c}_K\}$ of C such that $\mathbf{c}_i \in \mathbf{F}_q^N$ for all $i = 1, \ldots, K$.

6.21 Show that

$$\dim_{\mathbf{F}_q}(\text{Tr}_{\mathbf{F}_{q^m}/\mathbf{F}_q}(C)) \geq \dim_{\mathbf{F}_{q^m}}(C)$$

for any linear code C over $\mathbf{F}_{q^m}$.

6.22 (i) Show that, for a polynomial $f(x) \in \mathbf{F}_q[x]$ and an element $\alpha \in \mathbf{F}_{q^2}$, one has

$$(f(\alpha))^q + f(\alpha) \in \mathbf{F}_q \quad \text{and} \quad (f(\alpha))^{q+1} \in \mathbf{F}_q.$$

   (ii) Show that the set

$$S_m = \{x^{i(q+1)}(x^{jq} + x^j) : i(q + 1) + jq \leq (q + 1)m\}$$

has $(m + 1)(m + 2)/2$ elements. Moreover, show that the vector space $V_m =< S_m >$ spanned by S_m over $\mathbf{F}_q$ has dimension $|S_m| = (m + 1)(m + 2)/2$.

(iii) For an element $\beta \in \mathbf{F}_{q^2} \backslash \mathbf{F}_q$, we have $\beta^q \neq \beta$. Thus, we can label all the elements of $\mathbf{F}_{q^2} \backslash \mathbf{F}_q$ as follows:

$$\mathbf{F}_{q^2} \backslash \mathbf{F}_q = \{\beta_1, \beta_1^q, \ldots, \beta_n, \beta_n^q\},$$

where $n = (q^2 - q)/2$. Show that, for $m < q - 1$, the code

$$C_m = \{(g(\beta_1), \ldots, g(\beta_n)) : g \in V_m\}$$

is an $[n, (m+1)(m+2)/2, d]$-linear code over $\mathbf{F}_q$ with $d \geq n - m(q+1)/2$.

6.23 Let A be an $[N, K, D]$-linear code over $\mathbf{F}_{q^m}$ and let B be an $[n, m, d]$-linear code over $\mathbf{F}_q$. We set up an $\mathbf{F}_q$-linear transformation ϕ between $\mathbf{F}_{q^m}$ and B such that ϕ is bijective. We extend the map ϕ and obtain a map

$$\phi^* : \quad \mathbf{F}_{q^m}^N \to \mathbf{F}_q^{nN}, \qquad (v_1, \ldots, v_N) \mapsto (\phi(v_1), \ldots, \phi(v_N)).$$

Show that the code

$$(B^\perp)^N := \{(\mathbf{c}_1, \ldots, \mathbf{c}_N) : \mathbf{c}_i \in B^\perp\}$$

is contained in $(\phi^*(A))^\perp$.

6.24 Let C be a q^m-ary linear code. Show that

$$\dim_{\mathbf{F}_{q^m}}(C) - (m-1)(n - \dim_{\mathbf{F}_{q^m}}(C)) \leq \dim_{\mathbf{F}_q}(C|_{\mathbf{F}_q}) \leq \dim_{\mathbf{F}_{q^m}}(C)$$

and

$$\dim_{\mathbf{F}_{q^m}}(C) \leq \dim_{\mathbf{F}_q}(\mathrm{Tr}_{\mathbf{F}_{q^m}/\mathbf{F}_q}(C)) \leq m \cdot \dim_{\mathbf{F}_{q^m}}(C).$$

6.25 Let C be a q^m-ary linear code of length n and let U be an $\mathbf{F}_{q^m}$-subspace of C with the additional property $U^q \subseteq C$, where

$$U^q = \{(u_1^q, \ldots, u_n^q) : (u_1, \ldots, u_n) \in U\}.$$

Show that

$$\dim_{\mathbf{F}_q}(\mathrm{Tr}_{\mathbf{F}_{q^m}/\mathbf{F}_q}(C)) \leq m(\dim_{\mathbf{F}_{q^m}}(C) - \dim_{\mathbf{F}_{q^m}}(U)) + \dim_{\mathbf{F}_q}(U|_{\mathbf{F}_q}).$$

6.26 Let C be a q^m-ary linear code of length n and let V be an $\mathbf{F}_{q^m}$-subspace of $C^\perp$ with the additional property $V^q \subseteq C^\perp$. Show that

$$\dim_{\mathbf{F}_q}(C|_{\mathbf{F}_q}) \geq \dim_{\mathbf{F}_{q^m}}(C) - (m-1)(n - \dim_{\mathbf{F}_{q^m}}(C) - \dim_{\mathbf{F}_{q^m}}(V)).$$

6.27 Let C be a q^m-ary linear code of length n. Show that the following three conditions are equivalent:
   (i) $C^q = C$;
   (ii) $\dim_{\mathbf{F}_q}(C|_{\mathbf{F}_q}) = \dim_{\mathbf{F}_{q^m}}(C)$;
   (iii) $\mathrm{Tr}_{\mathbf{F}_{q^m}/\mathbf{F}_q}(C) = C|_{\mathbf{F}_q}$.

6.28 (Alphabet extension.) Let s and r be two integers such that $s \geq r > 1$. We embed an alphabet A of cardinality r into an alphabet B of cardinality s. For an (n, M, d)-code C over A, consider an embedding

$$C \hookrightarrow A^n \hookrightarrow B^n.$$

The code C can be viewed as a subset of B^n and therefore a code over B. Show that the code C still has parameters (n, M, d) when viewed as a code over B.

6.29 Let r and s be two integers bigger than 1. Let C_1 be an (n, M_1, d_1)-code over $\mathbf{Z}_r$, and let C_2 be an (n, M_2, d_2)-code over $\mathbf{Z}_s$. We embed $\mathbf{Z}_r$ ($\mathbf{Z}_s$, respectively) into $\mathbf{Z}_{rs}$ by mapping $(i \pmod r)) \in \mathbf{Z}_r$ ($(i \pmod s)) \in \mathbf{Z}_s$, respectively) to $(i \pmod{rs})) \in \mathbf{Z}_{rs}$. Then both C_1 and C_2 can be viewed as codes over $\mathbf{Z}_{rs}$. Show that the code

$$C_1 + rC_2 := \left\{ \mathbf{a} + r\mathbf{b} \in \mathbf{Z}_{rs}^n : \mathbf{a} \in C_1, \mathbf{b} \in C_2 \right\}$$

is an $(n, M_1 M_2, \min\{d_1, d_2\})$-code over $\mathbf{Z}_{rs}$.

6.30 (Alphabet restriction.) Let s and r be two integers such that $s \geq r > 1$. We embed an alphabet A of cardinality r into $\mathbf{Z}_s$. For an (n, M, d)-code C over $\mathbf{Z}_s$, consider all the s^n shifts

$$C_{\mathbf{v}} := \{\mathbf{v} + \mathbf{c} : \mathbf{c} \in C\}$$

for all $\mathbf{v} \in \mathbf{Z}_s^n$. Show that there exists a vector $\mathbf{v}_0 \in \mathbf{Z}_s^n$ such that the intersection $C_{\mathbf{v}_0} \cap A^n$ is an r-ary (n, M', d')-code with $M' \geq M(r/s)^n$, and $d' \geq d$.

7 Cyclic codes

In the previous chapters, we concentrated mostly on linear codes because they have algebraic structures. These structures simplify the study of linear codes. For example, a linear code can be described by its generator or parity-check matrices; the minimum distance is determined by the Hamming weight, etc. However, we have to introduce more structures besides linearity in order for codes to be implemented easily. For the sake of easy encoding and decoding, one naturally requires a cyclic shift of a codeword in a code C to be still a codeword of C. This requirement looks like a combinatorial structure. Fortunately, this structure can be converted into an algebraic one. Moreover, we will see that a cyclic code of length n is totally determined by a polynomial of degree less than n.

Cyclic codes were first studied by Prange [17] in 1957. Since then, algebraic coding theorists have made great progress in the study of cyclic codes for both random-error correction and burst-error correction. Many important classes of codes are among cyclic codes, such as the Hamming codes, Golay codes and the codes in Chapters 8 and 9.

We first define cyclic codes in this chapter, and then discuss their algebraic structure and other properties. In the final two sections, a decoding algorithm and burst-error-correcting codes are studied.

7.1 Definitions

Definition 7.1.1 A subset S of $\mathbf{F}_q^n$ is *cyclic* if $(a_{n-1}, a_0, a_1, \ldots, a_{n-2}) \in S$ whenever $(a_0, a_1, \ldots, a_{n-1}) \in S$. A linear code C is called a *cyclic code* if C is a cyclic set.

The word $(u_{n-r}, \ldots, u_{n-1}, u_0, u_1, \ldots, u_{n-r-1})$ is said to be obtained from the word $(u_0, \ldots, u_{n-1}) \in \mathbf{F}_q^n$ by cyclically shifting r positions.

It is easy to verify that the dual code of a cyclic code is also a cyclic code (see Exercise 7.2.).

Example 7.1.2 The sets

$$\{(0, 1, 1, 2), (2, 0, 1, 1), (1, 2, 0, 1), (1, 1, 2, 0)\} \subset \mathbf{F}_3^4, \quad \{11111\} \subset \mathbf{F}_2^5.$$

are cyclic sets, but they are not cyclic codes since they are not linear spaces.

Example 7.1.3 The following codes are cyclic codes:

(i) three trivial codes $\{\mathbf{0}\}$, $\{\lambda \cdot \mathbf{1} : \lambda \in \mathbf{F}_q\}$ and $\mathbf{F}_q^n$;
(ii) the binary $[3, 2, 2]$-linear code $\{000, 110, 101, 011\}$;
(iii) the simplex code $S(3, 2) = \{0000000, 1011100, 0101110, 0010111,$
        $1110010, 0111001, 1001011, 1100101\}.$

In order to convert the combinatorial structure of cyclic codes into an algebraic one, we consider the following correspondence:

$$\pi : \mathbf{F}_q^n \longrightarrow \mathbf{F}_q[x]/(x^n - 1), \quad (a_0, a_1, \ldots, a_{n-1}) \mapsto a_0 + a_1 x + \cdots + a_{n-1} x^{n-1}. \tag{7.1}$$

Then π is an $\mathbf{F}_q$-linear transformation of vector spaces over $\mathbf{F}_q$. From now on, we will sometimes identify $\mathbf{F}_q^n$ with $\mathbf{F}_q[x]/(x^n - 1)$, and a vector $\mathbf{u} = (u_0, u_1, \ldots, u_{n-1})$ with the polynomial $u(x) = \sum_{i=0}^{n-1} u_i x^i$. From Theorem 3.2.6, we know that $\mathbf{F}_q[x]/(x^n - 1)$ is a ring (but not a field unless $n = 1$). Thus, we have a multiplicative operation besides the addition in $\mathbf{F}_q^n$.

Example 7.1.4 Consider the cyclic code $C = \{000, 110, 101, 011\}$; then $\pi(C) = \{0, 1 + x, 1 + x^2, x + x^2\} \subset \mathbf{F}_2[x]/(x^3 - 1)$.

Now we introduce an important notion in the study of cyclic codes.

Definition 7.1.5 Let R be a ring. A nonempty subset I of R is called an *ideal* if

(i) both $a + b$ and $a - b$ belong to I, for all $a, b \in I$;
(ii) $r \cdot a \in I$, for all $r \in R$ and $a \in I$.

Example 7.1.6 In the ring $\mathbf{F}_2[x]/(x^3 - 1)$, the subset

$$I := \{0, 1 + x, x + x^2, 1 + x^2\}$$

is an ideal.

Example 7.1.7 (i) In the ring $\mathbf{Z}$ of integers, all the even integers form an ideal.

(ii) For a fixed positive integer m, all the integers divisible by m form an ideal of $\mathbf{Z}$.

(iii) In the polynomial ring $\mathbf{F}_q[x]$, for a given nonzero polynomial $f(x)$, all the polynomials divisible by $f(x)$ form an ideal.

(iv) In the ring $\mathbf{F}_q[x]/(x^n-1)$, for a divisor $g(x)$ of x^n-1, all the polynomials divisible by $g(x)$ form an ideal.

Definition 7.1.8 An ideal I of a ring R is called a *principal ideal* if there exists an element $g \in I$ such that $I =< g >$, where

$$< g >:= \{gr : r \in R\}.$$

The element g is called a *generator* of I and I is said to be generated by g.

A ring R is called a *principal ideal ring* if every ideal of R is principal.

Note that generators of a principal ideal may not be unique.

Example 7.1.9 In Example 7.1.6, the ideal I is principal. In fact, $I = < 1 + x >$. Note that

$$0 \cdot (1+x) = 1 + x^3 = 0 = (1 + x + x^2)(1 + x),$$
$$1 \cdot (1+x) = 1 + x = (x + x^2)(1 + x),$$
$$x \cdot (1+x) = x + x^2 = (1 + x^2)(1 + x),$$
$$x^2 \cdot (1+x) = 1 + x^2 = (1 + x)(1 + x).$$

Theorem 7.1.10 *The rings* $\mathbf{Z}$, $\mathbf{F}_q[x]$ *and* $\mathbf{F}_q[x]/(x^n-1)$ *are all principal ideal rings.*

Proof. Let I be an ideal of $\mathbf{Z}$. If $I = \{0\}$, then $I =< 0 >$ is a principal ideal. Assume that $I \neq \{0\}$ and let m be the smallest positive integer in I. Let a be any element of I. By the division algorithm, we have

$$a = qm + r \tag{7.2}$$

for some integers q and $0 \le r \le m - 1$. The equality (7.2) implies that r is also an element of I since $r = a - qm$. This forces $r = 0$ by the choice of m. Hence, $I =< m >$. This shows that $\mathbf{Z}$ is a principal ideal ring.

Using the same arguments, we can easily show that $\mathbf{F}_q[x]$ is also a principal ideal ring.

Essentially the same method can be employed for the case $\mathbf{F}_q[x]/(x^n-1)$. Since this case is crucial for this chapter, we repeat the arguments. The zero ideal is obviously principal. We choose a nonzero polynomial $g(x)$ of a nonzero ideal J with the lowest degree. For any polynomial $f(x)$ of J, we have

$$f(x) = s(x)g(x) + r(x)$$

for some polynomials $s(x), r(x) \in \mathbf{F}_q[x]$ with $\deg(r(x)) < \deg(g(x))$. This forces $r(x) = 0$, since $r(x) = f(x) - s(x)g(x) \in J$ and $g(x)$ has the lowest degree among the nonzero polynomials of J. Hence, $J = < g(x) >$, and the desired result follows. $\square$

7.2 Generator polynomials

The reason for defining ideals in the preceding section is the following result connecting ideals and cyclic codes.

Theorem 7.2.1 *Let π be the linear map defined in (7.1). Then a nonempty subset C of $\mathbf{F}_q^n$ is a cyclic code if and only if $\pi(C)$ is an ideal of $\mathbf{F}_q[x]/(x^n - 1)$.*

Proof. Suppose that $\pi(C)$ is an ideal of $\mathbf{F}_q[x]/(x^n - 1)$. Then, for any $\alpha, \beta \in \mathbf{F}_q \subset \mathbf{F}_q[x]/(x^n-1)$ and $\mathbf{a}, \mathbf{b} \in C$, we have $\alpha\pi(\mathbf{a}), \beta\pi(\mathbf{b}) \in \pi(C)$ by Definition 7.1.5(ii). Thus by Definition 7.1.5(i), $\alpha\pi(\mathbf{a}) + \beta\pi(\mathbf{b})$ is an element of $\pi(C)$; i.e., $\pi(\alpha\mathbf{a} + \beta\mathbf{b}) \in \pi(C)$, hence $\alpha\mathbf{a} + \beta\mathbf{b}$ is a codeword of C. This shows that C is a linear code.

Now let $\mathbf{c} = (c_0, c_1, \ldots, c_{n-1})$ be a codeword of C. The polynomial

$$\pi(\mathbf{c}) = c_0 + c_1 x + \cdots + c_{n-2}x^{n-2} + c_{n-1}x^{n-1}$$

is an element of $\pi(C)$. Since $\pi(C)$ is an ideal, the element

$$\begin{aligned}
x\pi(\mathbf{c}) &= c_0 x + c_1 x^2 + \cdots + c_{n-2}x^{n-1} + c_{n-1}x^n \\
&= c_{n-1} + c_0 x + c_1 x^2 + \cdots + c_{n-2}x^{n-1} + c_{n-1}(x^n - 1) \\
&= c_{n-1} + c_0 x + c_1 x^2 + \cdots + c_{n-2}x^{n-1} \\
&\quad (\text{as } x^n - 1 = 0 \text{ in } \mathbf{F}_q[x]/(x^n - 1))
\end{aligned}$$

is in $\pi(C)$; i.e., $(c_{n-1}, c_0, c_1, \ldots, c_{n-2})$ is a codeword of C. This means that C is cyclic.

Conversely, suppose that C is a cyclic code. Then it is clear that (i) of Definition 7.1.5 is satisfied for $\pi(C)$. For any polynomial

$$f(x) = f_0 + f_1 x + \cdots + f_{n-2}x^{n-2} + f_{n-1}x^{n-1} = \pi(f_0, f_1, \ldots, f_{n-1})$$

of $\pi(C)$ with $(f_0, f_1, \ldots, f_{n-1}) \in C$, the polynomial

$$xf(x) = f_{n-1} + f_0 x + f_1 x^2 + \cdots + f_{n-2} x^{n-1}$$

is also an element of $\pi(C)$ since C is cyclic. Thus, $x^2 f(x) = x(xf(x))$ is an element of $\pi(C)$. By induction, we know that $x^i f(x)$ belongs to $\pi(C)$ for all $i \geq 0$. Since C is a linear code and π is a linear transformation, $\pi(C)$ is a linear space over $\mathbf{F}_q$. Hence, for any $g(x) = g_0 + g_1 x + \cdots + g_{n-1} x^{n-1} \in \mathbf{F}_q[x]/(x^n - 1)$, the polynomial

$$g(x)f(x) = \sum_{i=0}^{n-1} g_i(x^i f(x))$$

is an element of $\pi(C)$. Therefore, $\pi(C)$ is an ideal of $\mathbf{F}_q[x]/(x^n - 1)$ since (ii) of Definition 7.1.5 is also satisfied. □

Example 7.2.2 (i) The code $C = \{(0, 0, 0), (1, 1, 1), (2, 2, 2)\}$ is a ternary cyclic code. The corresponding ideal in $\mathbf{F}_3[x]/(x^3 - 1)$ is $\pi(C) = \{0, 1 + x + x^2, 2 + 2x + 2x^2\}$.

(ii) The set $I = \{0, 1 + x^2, x + x^3, 1 + x + x^2 + x^3\}$ is an ideal in $\mathbf{F}_2[x]/(x^4 - 1)$. The corresponding cyclic code is $\pi^{-1}(I) = \{0000, 1010, 0101, 1111\}$.

(iii) The trivial cyclic codes $\{\mathbf{0}\}$ and $\mathbf{F}_q^n$ correspond to the trivial ideals $\{0\}$ and $\mathbf{F}_q[x]/(x^n - 1)$, respectively.

Theorem 7.2.3 Let I be a nonzero ideal in $\mathbf{F}_q[x]/(x^n - 1)$ and let $g(x)$ be a nonzero monic polynomial of the least degree in I. Then $g(x)$ is a generator of I and divides $x^n - 1$.

Proof. For the first part, we refer to the proof of Theorem 7.1.10.
Consider the division algorithm

$$x^n - 1 = s(x)g(x) + r(x)$$

with $\deg(r(x)) < \deg(g(x))$. Hence,

$$r(x) = (x^n - 1) - s(x)g(x)$$

is an element of I (note that $x^n - 1$ is the zero element of $\mathbf{F}_q[x]/(x^n - 1)$). This implies that $r(x) = 0$ since $g(x)$ has the lowest degree. Hence, $g(x)$ is a divisor of $x^n - 1$. □

Example 7.2.4 In Example 7.2.2(i), the polynomial $1 + x + x^2$ is of the least degree. It divides $x^3 - 1$. In Example 7.2.2(ii), the polynomial $1 + x^2$ is of the least degree. It divides $x^4 - 1$.
For the code $\mathbf{F}_q^n$, the polynomial 1 is of the least degree.

By Theorem 7.1.10, we know that every ideal in $\mathbf{F}_q[x]/(x^n - 1)$ is principal, thus a cyclic code C is determined by any of the generators of $\pi(C)$. Usually, there is more than one generator for an ideal of $\mathbf{F}_q[x]/(x^n - 1)$. The following result shows that the generator satisfying certain additional properties is unique.

Theorem 7.2.5 *There is a unique monic polynomial of the least degree in every nonzero ideal I of $\mathbf{F}_q[x]/(x^n - 1)$. (By Theorem 7.2.3, it is a generator of I.)*

Proof. Let $g_i(x)$, $i = 1, 2$, be two distinct monic generators of the least degree of the ideal I. Then, a suitable scalar multiple of $g_1(x) - g_2(x)$ is a nonzero monic polynomial of smaller degree in I. It is a contradiction. $\square$

From the above result, the following definition makes sense.

Definition 7.2.6 The unique monic polynomial of the least degree of a nonzero ideal I of $\mathbf{F}_q[x]/(x^n - 1)$ is called the *generator polynomial* of I. For a cyclic code C, the generator polynomial of $\pi(C)$ is also called the *generator polynomial* of C.

Example 7.2.7 (i) The generator polynomial of the cyclic code {000, 110, 011, 101} is $1 + x$.

(ii) The generator polynomial of the simplex code in Example 7.1.3(iii) is $1 + x^2 + x^3 + x^4$.

Theorem 7.2.8 *Each monic divisor of $x^n - 1$ is the generator polynomial of some cyclic code in $\mathbf{F}_q^n$.*

Proof. Let $g(x)$ be a monic divisor of $x^n - 1$ and let I be the ideal $< g(x) >$ of $\mathbf{F}_q[x]/(x^n - 1)$ generated by $g(x)$. Let C be the corresponding cyclic code. Assume that $h(x)$ is the generator polynomial of C. Then there exists a polynomial $b(x)$ such that

$$h(x) \equiv g(x)b(x) \pmod{x^n - 1}.$$

Thus, $g(x)$ is a divisor of $h(x)$. Hence, $g(x)$ is the same as $h(x)$ since $h(x)$ has the least degree and is monic. $\square$

From Theorems 7.2.5 and 7.2.8, we obtain the following result.

Corollary 7.2.9 *There is a one-to-one correspondence between the cyclic codes in $\mathbf{F}_q^n$ and the monic divisors of $x^n - 1 \in \mathbf{F}_q[x]$.*

Remark 7.2.10 The polynomials 1 and $x^n - 1$ correspond to $\mathbf{F}_q^n$ and {0}, respectively.

Example 7.2.11 In order to find all binary cyclic codes of length 6, we factorize the polynomial $x^6 - 1 \in \mathbf{F}_2[x]$:

$$x^6 - 1 = (1 + x)^2 (1 + x + x^2)^2.$$

List all the monic divisors of $x^6 - 1$:

$$1, \qquad 1 + x, \qquad 1 + x + x^2,$$
$$(1 + x)^2, \qquad (1 + x)(1 + x + x^2), \qquad (1 + x)^2(1 + x + x^2),$$
$$(1 + x + x^2)^2, \qquad (1 + x)(1 + x + x^2)^2, \qquad 1 + x^6.$$

Thus, there are nine binary cyclic codes of length 6 altogether. Based on the map π, we can easily write down all these cyclic codes. For instance, the cyclic code corresponding to the polynomial $(1 + x + x^2)^2$ is

$$\{000000, 101010, 010101, 111111\}.$$

From the above example, we find that the number of cyclic codes of length n can be determined if we know the factorization of $x^n - 1$. We have the following result.

Theorem 7.2.12 *Let* $x^n - 1 \in \mathbf{F}_q[x]$ *have the factorization*

$$x^n - 1 = \prod_{i=1}^{r} p_i^{e_i}(x),$$

where $p_1(x), p_2(x), \ldots, p_r(x)$ *are distinct monic irreducible polynomials and* $e_i \geq 1$ *for all* $i = 1, 2, \ldots, r$. *Then there are* $\prod_{i=1}^{r}(e_i + 1)$ *cyclic codes of length n over* $\mathbf{F}_q$.

The proof of Theorem 7.2.12 follows from Corollary 7.2.9 by counting the number of monic divisors of $x^n - 1$.

Example 7.2.13 Using Theorem 3.4.11, we can factorize the polynomial $x^n - 1$, and thus the number of cyclic codes of length n can be determined by Theorem 7.2.12.

Tables 7.1 and 7.2 show the factorization of $x^n - 1$ and the number of q-ary cyclic codes of length n, for $1 \leq n \leq 10$ and $q = 2, 3$.

Since a cyclic code is totally determined by its generator polynomial, all the parameters of the code are also determined by the generator polynomial. The following result gives the dimension in terms of the generator polynomial.

Table 7.1. Binary cyclic codes of length up to 10.

n	Factorization of $x^n - 1$	No. of cyclic codes
1	$1 + x$	2
2	$(1 + x)^2$	3
3	$(1 + x)(1 + x + x^2)$	4
4	$(1 + x)^4$	5
5	$(1 + x)(1 + x + x^2 + x^3 + x^4)$	4
6	$(1 + x)^2(1 + x + x^2)^2$	9
7	$(1 + x)(1 + x^2 + x^3)(1 + x + x^3)$	8
8	$(1 + x)^8$	9
9	$(1 + x)(1 + x + x^2)(1 + x^3 + x^6)$	8
10	$(1 + x)^2(1 + x + x^2 + x^3 + x^4)^2$	9

Theorem 7.2.14 *Let $g(x)$ be the generator polynomial of an ideal of $\mathbf{F}_q[x]/$ $(x^n - 1)$. Then the corresponding cyclic code has dimension k if the degree of $g(x)$ is $n - k$.*

Proof. For two polynomials $c_1(x) \neq c_2(x)$ with $\deg(c_i(x)) \leq k - 1$ ($i = 1, 2$), we have clearly that $g(x)c_1(x) \not\equiv g(x)c_2(x) \pmod{x^n - 1}$. Hence, the set

$$A := \{g(x)c(x) : c(x) \in \mathbf{F}_q[x]/(x^n - 1), \ \deg(c(x)) \leq k - 1\}$$

has q^k elements and is a subset of the ideal $< g(x) >$. On the other hand, for any codeword $g(x)a(x)$ with $a(x) \in \mathbf{F}_q[x]/(x^n - 1)$, we write

$$a(x)g(x) = u(x)(x^n - 1) + v(x) \tag{7.3}$$

with $\deg(v(x)) < n$. By (7.3), we have that $v(x) = a(x)g(x) - u(x)(x^n - 1)$. Hence, $g(x)$ divides $v(x)$. Write $v(x) = g(x)b(x)$ for some polynomial $b(x)$. Then $\deg(b(x)) < k$, so $v(x)$ is in A. This shows that A is the same as $< g(x) >$. Hence, the dimension of the code is $\log_q |A| = k$. $\qquad \square$

Example 7.2.15 (i) Based on the factorization: $x^7 - 1 = (1 + x)(1 + x^2 + x^3)(1 + x + x^3) \in \mathbf{F}_2[x]$, we know that there are only two binary [7, 3]-cyclic codes:

$$< (1 + x)(1 + x^2 + x^3) > \ = \ \{0000000, 1110100, 0111010, 0011101,$$
$$1001110, 0100111, 1010011, 1101001\}$$

Table 7.2. Ternary cyclic codes of length up to 10.

n	Factorization of $x^n - 1$	No. of cyclic codes
1	$2 + x$	2
2	$(2 + x)(1 + x)$	4
3	$(2 + x)^3$	4
4	$(2 + x)(1 + x)(1 + x^2)$	8
5	$(2 + x)(1 + x + x^2 + x^3 + x^4)$	4
6	$(2 + x)^3(1 + x)^3$	16
7	$(2 + x)(1 + x + x^2 + x^3 + x^4 + x^5 + x^6)$	4
8	$(2 + x)(1 + x)(1 + x^2)(2 + x + x^2)$ $(2 + 2x + x^2)$	32
9	$(2 + x)^9$	10
10	$(2 + x)(1 + x)(1 + x + x^2 + x^3 + x^4)$ $(1 + 2x + x^2 + 2x^3 + x^4)$	16

and

$$< (1 + x)(1 + x + x^3) > \; = \; \{0000000, 1011100, 0101110, 0010111,$$
$$1001011, 1100101, 1110010, 0111001\}.$$

(ii) Based on the factorization: $x^7 - 1 = (2 + x)(1 + x + x^2 + x^3 + x^4 + x^5 + x^6) \in \mathbf{F}_3[x]$, we do not have any ternary $[7, 2]$-cyclic codes.

7.3   Generator and parity-check matrices

In the previous section, we showed that a cyclic code is totally determined by its generator polynomial. Hence, such a code should also have generator matrices determined by this polynomial. Indeed, we have the following result.

Theorem 7.3.1 *Let* $g(x) = g_0 + g_1 x + \cdots + g_{n-k} x^{n-k}$ *be the generator polynomial of a cyclic code* C *in* $\mathbf{F}_q^n$ *with* $\deg(g(x)) = n - k$. *Then the matrix*

$$G = \begin{pmatrix} g(x) \\ xg(x) \\ \cdot \\ \cdot \\ \cdot \\ x^{k-1}g(x) \end{pmatrix} = \begin{pmatrix} g_0 & g_1 & \cdot & \cdot & \cdot & g_{n-k} & 0 & 0 & 0 & \cdot & \cdot & 0 \\ 0 & g_0 & g_1 & \cdot & \cdot & & g_{n-k} & 0 & 0 & \cdot & \cdot & 0 \\ \cdot & & & & & & & & & & & \\ \cdot & & & & & & & & & & & \\ \cdot & & & & & & & & & & & \\ 0 & 0 & \cdot & \cdot & \cdot & g_0 & g_1 & \cdot & \cdot & \cdot & \cdot & g_{n-k} \end{pmatrix}$$

is a generator matrix of C *(note that we identify a vector with a polynomial).*

Proof. It is sufficient to show that $g(x), xg(x), \ldots, x^{k-1}g(x)$ form a basis of C. It is clear that they are linearly independent over $\mathbf{F}_q$. By Theorem 7.2.14, we know that $\dim(C) = k$. The desired result follows.   □

Example 7.3.2 Consider the binary $[7, 4]$-cyclic code with generator polynomial $g(x) = 1 + x^2 + x^3$. Then this code has a generator matrix

$$G = \begin{pmatrix} g(x) \\ xg(x) \\ x^2 g(x) \\ x^3 g(x) \end{pmatrix} = \begin{pmatrix} 1 & 0 & 1 & 1 & 0 & 0 & 0 \\ 0 & 1 & 0 & 1 & 1 & 0 & 0 \\ 0 & 0 & 1 & 0 & 1 & 1 & 0 \\ 0 & 0 & 0 & 1 & 0 & 1 & 1 \end{pmatrix}.$$

This generator matrix is not in standard form. If the fourth row is added to the second row and the sum of the last two rows is added to the first row, we form a generator matrix in standard form:

$$G' = \begin{pmatrix} 1 & 0 & 0 & 0 & 1 & 0 & 1 \\ 0 & 1 & 0 & 0 & 1 & 1 & 1 \\ 0 & 0 & 1 & 0 & 1 & 1 & 0 \\ 0 & 0 & 0 & 1 & 0 & 1 & 1 \end{pmatrix}.$$

Thus, a parity-check matrix is easily obtained from G' by Algorithm 4.3.

From the above example, we know that parity-check matrices of a cyclic code can be obtained from its generator matrices by performing elementary row operations. However, since the dual code of a cyclic code C is also cyclic, we should be able to find a parity-check matrix from the generator polynomial of the dual code. The question then is to find the generator polynomial of the dual code $C^{\perp}$.

Definition 7.3.3 Let $h(x) = \sum_{i=0}^{k} a_i x^i$ be a polynomial of degree k $(a_k \neq 0)$ over $\mathbf{F}_q$. Define the *reciprocal polynomial* $h_R(x)$ of $h(x)$ by

$$h_R(x) := x^k h(1/x) = \sum_{i=0}^{k} a_{k-i} x^i.$$

Remark 7.3.4 If $h(x)$ is a divisor of $x^n - 1$, then so is $h_R(x)$.

Example 7.3.5 (i) For the polynomial $h(x) = 1 + 2x + 3x^5 + x^7 \in \mathbf{F}_5[x]$, the reciprocal of $h(x)$ is

$$\begin{aligned} h_R(x) &= x^7 h(1/x) \\ &= x^7 (1 + 2(1/x) + 3(1/x)^5 + (1/x)^7) \\ &= 1 + 3x^2 + 2x^6 + x^7. \end{aligned}$$

(ii) Consider the divisor $h(x) = 1 + x + x^3 \in \mathbf{F}_2[x]$ of $x^7 - 1$. Then $h_R(x) = 1 + x^2 + x^3$ is also a divisor of $x^7 - 1$.

Example 7.3.6 Let $g(x) = g_0 + g_1x + g_2x^2 + g_3x^3$ be the generator polynomial of a cyclic code C over $\mathbf{F}_q$ of length 4 and let $h(x) = (x^4 - 1)/g(x)$. Put $h(x) = h_0 + h_1x + h_2x^2 + h_3x^3$. Then $h_R(x) = (h_3 + h_2x + h_1x^2 + h_0x^3)/x^{3-k}$, where $k = \deg(h(x))$.

Consider the product

$$
\begin{aligned}
0 &\equiv g(x)h(x) \\
&\equiv (g_0 + g_1x + g_2x^2 + g_3x^3)(h_0 + h_1x + h_2x^2 + h_3x^3) \\
&\equiv g_0h_0 + (g_0h_1 + g_1h_0)x + (g_0h_2 + g_1h_1 + g_2h_0)x^2 + (g_0h_3 \\
&\quad + g_1h_2 + g_2h_1 + g_3h_0)x^3 + (g_1h_3 + g_2h_2 + g_3h_1)x^4 \\
&\quad + (g_2h_3 + g_3h_2)x^5 + g_3h_3x^6 \\
&\equiv (g_0h_0 + g_1h_3 + g_2h_2 + g_3h_1) + (g_0h_1 + g_1h_0 + g_2h_3 \\
&\quad + g_3h_2)x + (g_0h_2 + g_1h_1 + g_2h_0 + g_3h_3)x^2 + (g_0h_3 + g_1h_2 \\
&\quad + g_2h_1 + g_3h_0)x^3 \pmod{x^4 - 1}.
\end{aligned}
\tag{7.4}
$$

Thus, the coefficient of each power of x at the last step of (7.4) must be zero.

Put $\mathbf{b} = (h_3, h_2, h_1, h_0) \in \mathbf{F}_q^4$ and $\mathbf{g} = (g_0, g_1, g_2, g_3) \in \mathbf{F}_q^4$. Let $\mathbf{g}_i$ be the vector obtained from $\mathbf{g}$ by cyclically shifting i positions. By looking at the coefficient of x^3 in (7.4), we obtain

$$\mathbf{g}_0 \cdot \mathbf{b} = \mathbf{g} \cdot \mathbf{b} = g_0h_3 + g_1h_2 + g_2h_1 + g_3h_0 = 0.$$

By looking at the coefficients of the other powers of x in (7.4), we obtain $\mathbf{g}_i \cdot \mathbf{b} = 0$ for all $i = 0, 1, 2, 3$. Therefore, $\mathbf{b}$ is a codeword of $C^\perp$ since the set $\{\mathbf{g}_0, \mathbf{g}_1, \mathbf{g}_2, \mathbf{g}_3\}$ generates C by Theorem 7.3.1.

By cyclically shifting the vector $\mathbf{b} = (h_3, h_2, h_1, h_0)$ by $k + 1$ positions, we obtain the vector corresponding to $h_R(x)$. This implies that $h_R(x)$ is a codeword as $C^\perp$ is also a cyclic code.

Since $\deg(h_R(x)) = \deg(h(x)) = k$, the set $\{h_R(x), xh_R(x), \ldots, x^{n-k-1}h_R(x)\}$ is a basis of $C^\perp$. Hence, $C^\perp$ is generated by $h_R(x)$. Thus, the monic polynomial $h_0^{-1}h_R(x)$ is the generator polynomial of $C^\perp$ (note that $h_0 = h(0) \neq 0$ since $h_0g_0 = -1$).

It is clear that the above example can be easily generalized to any length n.

Theorem 7.3.7 Let $g(x)$ be the generator polynomial of a q-ary $[n, k]$-cyclic code C. Put $h(x) = (x^n - 1)/g(x)$. Then $h_0^{-1}h_R(x)$ is the generator polynomial of $C^\perp$, where h_0 is the constant term of $h(x)$.

Proof. Let $g(x) = \sum_{i=0}^{n-1} g_i x^i$ and let $h(x) = \sum_{i=0}^{n-1} h_i x^i$. Then

$$h_R(x) = (1/x^{n-k-1}) \sum_{i=0}^{n-1} h_{n-i-1} x^i,$$

where $k = \deg(h(x))$.

Consider the product

$$
\begin{aligned}
0 &\equiv g(x)h(x) \\
&\equiv (g_0 h_0 + g_1 h_{n-1} + \cdots + g_{n-1}h_1) + (g_0 h_1 + g_1 h_0 + \cdots + g_{n-1}h_2)x \\
&\quad + (g_0 h_2 + g_1 h_1 + \cdots + g_{n-1}h_3)x^2 + \cdots + (g_0 h_{n-1} + g_1 h_{n-2} \\
&\quad + \cdots + g_{n-1}h_0)x^{n-1} \pmod{x^n - 1}.
\end{aligned}
$$

Thus, the coefficient of each power of x in the last line of the above display must be zero. By looking at the coefficient of each power of x, we obtain $\mathbf{g}_i \cdot (h_{n-1}, h_{n-2}, \ldots, h_1, h_0) = 0$, for all $i = 0, 1, \ldots, n-1$, where $\mathbf{g}_i$ is the vector obtained from $(g_0, g_1, \ldots, g_{n-1})$ by cyclically shifting i positions. Therefore, $(h_{n-1}, h_{n-2}, \ldots, h_1, h_0)$ is a codeword of $C^\perp$ since $\{\mathbf{g}_0, \mathbf{g}_1, \ldots, \mathbf{g}_{n-1}\}$ generates C by Theorem 7.3.1.

By cyclically shifting the vector $(h_{n-1}, h_{n-2}, \ldots, h_1, h_0)$ by $k+1$ positions, we obtain the vector corresponding to $h_R(x)$. This implies that $h_R(x)$ is a codeword as $C^\perp$ is also a cyclic code.

Since $\deg(h_R(x)) = \deg(h(x)) = k$, the set $\{h_R(x), x h_R(x), \ldots, x^{n-k-1} h_R(x)\}$ is a basis of $C^\perp$. Hence, $C^\perp$ is generated by $h_R(x)$. Thus, the monic polynomial $h_0^{-1} h_R(x)$ is the generator polynomial of $C^\perp$. □

Definition 7.3.8 Let C be a q-ary cyclic code of length n. Put $h(x) = (x^n - 1)/g(x)$. Then, $h_0^{-1} h_R(x)$ is called the *parity-check polynomial* of C, where h_0 is the constant term of $h(x)$.

Corollary 7.3.9 *Let C be a q-ary $[n, k]$-cyclic code with generator polynomial $g(x)$. Put $h(x) = (x^n - 1)/g(x)$. Let $h(x) = h_0 + h_1 x + \cdots + h_k x^k$. Then the matrix*

$$
H = \begin{pmatrix} h_R(x) \\ x h_R(x) \\ \cdot \\ \cdot \\ \cdot \\ x^{n-k-1} h_R(x) \end{pmatrix} = \begin{pmatrix} h_k & h_{k-1} & \cdot & \cdot & \cdot & h_0 & 0 & 0 & 0 & \cdot & \cdot & 0 \\ 0 & h_k & h_{k-1} & \cdot & \cdot & \cdot & h_0 & 0 & 0 & \cdot & \cdot & 0 \\ \cdot & & & & & & & & & & & \\ \cdot & & & & & & & & & & & \\ \cdot & & & & & & & & & & & \\ 0 & 0 & \cdot & \cdot & \cdot & h_k & h_{k-1} & \cdot & \cdot & \cdot & \cdot & h_0 \end{pmatrix}
$$

is a parity-check matrix of C.

Proof. The result immediately follows from Theorems 7.3.1 and 7.3.7.   □

Example 7.3.10 Let C be the binary $[7, 4]$-cyclic code generated by $g(x) = 1 + x^2 + x^3$ as in Example 7.3.2. Put $h(x) = (x^7 - 1)/g(x) = 1 + x^2 + x^3 + x^4$. Then $h_R(x) = 1 + x + x^2 + x^4$ is the parity-check polynomial of C. Hence,

$$H = \begin{pmatrix} 1 & 1 & 1 & 0 & 1 & 0 & 0 \\ 0 & 1 & 1 & 1 & 0 & 1 & 0 \\ 0 & 0 & 1 & 1 & 1 & 0 & 1 \end{pmatrix}$$

is a parity-check matrix of C.

7.4 Decoding of cyclic codes

The decoding of cyclic codes consists of the same three steps as the decoding of linear codes: computing the syndrome; finding the syndrome corresponding to the error pattern; and correcting the errors. Because of the pleasing structure of cyclic codes, the three steps for cyclic codes are usually simpler. Cyclic codes have considerable algebraic and geometric properties. If these properties are properly used, simplicity in the decoding can be easily achieved.

From Corollary 7.3.9, for a cyclic code, we can easily produce a parity-check matrix of the form

$$H = (I_{n-k}|A) \tag{7.5}$$

by performing elementary row operations. Though parity-check matrices for a linear code are not unique, the parity-check matrix of the form in (7.5) is unique. All syndromes considered in this section are computed with respect to the parity-check matrix of the form in (7.5).

Theorem 7.4.1 Let $H = (I_{n-k}|A)$ be a parity-check matrix of a q-ary cyclic code C. Let $g(x)$ be the generator polynomial of C. Then the syndrome of a vector $\mathbf{w} \in F_q^n$ is equal to $(w(x) \pmod{g(x)})$; i.e., the principal remainder of $w(x)$ divided by $g(x)$ (note that here we identify a vector of F_q^n with a polynomial of $F_q[x]/(x^n - 1)$, and thus $w(x)$ is the corresponding polynomial of $\mathbf{w}$).

Proof. For each column vector of A, we associate a polynomial of degree at most $n - k - 1$ and write A as

$$A = (a_0(x), a_1(x), \dots, a_{k-1}(x)).$$

By Algorithm 4.3, we know that $G = (-A^T | I_k)$ is a generator matrix for C. Therefore, $x^{n-k+i} - a_i(x)$ is a codeword of C. Put $x^{n-k+i} - a_i(x) = q_i(x)g(x)$ for some $q_i(x) \in \mathbf{F}_q[x]/(x^n - 1)$; i.e.,

$$a_i(x) = x^{n-k+i} - q_i(x)g(x). \tag{7.6}$$

Suppose $w(x) = w_0 + w_1 x + \cdots + w_{n-1} x^{n-1}$. For the syndrome $\mathbf{s} = \mathbf{w} H^T$ of $\mathbf{w}$, the corresponding polynomial $s(x)$ is

$$
\begin{aligned}
s(x) &= w_0 + w_1 x + \cdots + w_{n-k-1} x^{n-k-1} + w_{n-k} a_0(x) + \cdots + w_{n-1} a_{k-1}(x) \\
&= \sum_{i=0}^{n-k-1} w_i x^i + \sum_{j=0}^{k-1} w_{n-k+j}(x^{n-k+j} - q_j(x)g(x)) \quad \text{(by (7.6))} \\
&= \sum_{i=0}^{n-1} w_i x^i - \left(\sum_{j=0}^{k-1} w_{n-k+j} q_j(x) \right) g(x) \\
&\equiv w(x) \pmod{g(x)}.
\end{aligned}
$$

As the polynomial $s(x)$ has degree at most $n - k - 1$, the desired result follows. $\square$

Example 7.4.2 Consider the binary $[7, 4, 3]$-Hamming code with the generator polynomial $g(x) = 1 + x^2 + x^3$. Then, by performing elementary row operations from the matrix in Example 7.3.10, we obtain a parity-check matrix $H = (I_3 | A)$, where A is the matrix

$$A = \begin{pmatrix} 1 & 1 & 1 & 0 \\ 0 & 1 & 1 & 1 \\ 1 & 1 & 0 & 1 \end{pmatrix}.$$

For the word $\mathbf{w} = 0110110$, the syndrome is $\mathbf{s} = \mathbf{w} H^T = 010$. On the other hand,

$$w(x) = x + x^2 + x^4 + x^5 = x + x^2 g(x).$$

Thus, the remainder $(w(x) \pmod{g(x)})$ is x, which corresponds to the word 010.

Theorem 7.4.1 shows that the syndrome of a received word $w(x)$ can be determined by the remainder $s(x) = (w(x) \pmod{g(x)})$. Hence, $w(x) - s(x)$ is a codeword.

Corollary 7.4.3 *Let $g(x)$ be the generator polynomial of a cyclic code C. For a received word $w(x)$, if the remainder $s(x)$ of $w(x)$ divided by $g(x)$ has weight* *less than or equal to $\lfloor (d(C)-1)/2 \rfloor$, then $s(x)$ is the error pattern of $w(x)$; i.e., $w(x)$ is decoded to $w(x) - s(x)$ by MLD.*

Proof. By Theorem 7.4.1, we know that $w(x)$ and $s(x)$ are in the same coset. Furthermore, $s(x)$ is a coset leader by Exercise 4.44 since wt$(s(x)) \le \lfloor (d(C)-1)/2 \rfloor$. The desired result follows. $\qquad\square$

Example 7.4.4 As in Example 7.4.2, the remainder of $w(x) = x + x^2 + x^4 + x^5$ divided by $g(x) = 1 + x^2 + x^3$ is x. Therefore, $w(x)$ is decoded to $w(x) - x = x^2 + x^4 + x^5 = 0010110$. If the word $w_1(x) = 1 + x^2 + x^3 + x^4$ is received, then the remainder $(w_1(x) \pmod{g(x)})$ is $1 + x + x^2$. In this case, we can use syndrome decoding to obtain the codeword $w_1(x) - x^4 = 1 + x^2 + x^3 = 1011000$ as the word 0000100 is the coset leader for the coset in which $w_1(x)$ lies.

From the above example, we see that, for some received words we can directly decode by throwing away the remainder from the words. However, for other words we have to use syndrome decoding. Because of the algebraic and geometric properties of cyclic codes, we can simplify the syndrome decoding for some received words. In the rest of this section, we will describe the so-called *error trapping* decoding.

Lemma 7.4.5 *Let C be a q-ary $[n, k]$-cyclic code with generator polynomial $g(x)$. Let $s(x) = \sum_{i=0}^{n-k-1} s_i x^i$ be the syndrome of $w(x)$. Then the syndrome of the cyclic shift $xw(x)$ is equal to $xs(x) - s_{n-k-1}g(x)$.*

Proof. By Theorem 7.4.1, it is sufficient to show that $xs(x) - s_{n-k-1}g(x)$ is the remainder of $xw(x)$ divided by $g(x)$. Let $w(x) = q(x)g(x) + s(x)$. Then

$$xw(x) = xq(x)g(x) + xs(x) = (xq(x) + s_{n-k-1})g(x) + (xs(x) - s_{n-k-1}g(x)).$$

The desired result follows as $\deg(xs(x) - s_{n-k-1}g(x)) < n - k = \deg(g(x))$. $\qquad\square$

Remark 7.4.6 The syndrome of the cyclic shift $x^i w(x)$ of a word $w(x)$ can be computed through the syndrome of the cyclic shift $x^{i-1} w(x)$. Thus, the syndromes of $w(x), xw(x), x^2 w(x), \ldots,$ can be computed inductively.

Example 7.4.7 As in Example 7.4.2, the syndrome of $w(x) = x + x^2 + x^4 + x^5$ is x, thus the syndromes of $xw(x)$ and $x^2 w(x)$ are $x \cdot x = x^2$ and $x \cdot x^2 - g(x) = 1 + x^2$, respectively.

Definition 7.4.8 A *cyclic run of* 0 *of length l* of an n-tuple is a succession of l cyclically consecutive zero components.

Example 7.4.9 (i) $\mathbf{e} = (1, 3, 0, 0, 0, 0, 0, 1, 0)$ has a cyclic run of 0 of length 5.

(ii) $\mathbf{e} = (0, 0, 1, 2, 0, 0, 0, 1, 0, 0)$ has a cyclic run of 0 of length 4.

Decoding algorithm for cyclic codes

Let C be a q-ary $[n, k, d]$-cyclic code with generator polynomial $g(x)$. Let $w(x)$ be a received word with an error pattern $e(x)$, where $\mathrm{wt}(e(x)) \le \lfloor (d-1)/2 \rfloor$ and $e(x)$ has a cyclic run of 0 of length at least k. The goal is to determine $e(x)$.

Step 1: Compute the syndromes of $x^i w(x)$, for $i = 0, 1, 2, \ldots$, and denote by $s_i(x)$ the syndrome $(x^i w(x) \pmod{g(x)})$.

Step 2: Find m such that the weight of the syndrome $s_m(x)$ for $x^m w(x)$ is less than or equal to $\lfloor (d-1)/2 \rfloor$.

Step 3: Compute the remainder $e(x)$ of $x^{n-m} s_m(x)$ divided by $x^n - 1$. Decode $w(x)$ to $w(x) - e(x)$.

Proof. First of all, we show the existence of such an m in Step 2. By the assumption, there exists an error pattern $e(x)$ such that $e(x)$ has a cyclic run of 0 of length at least k. Thus, there exists an integer $m \ge 0$ such that the cyclic shift of the error pattern $e(x)$ through m positions has all its nonzero components within the first $n - k$ positions. The cyclic shift of the error pattern $e(x)$ through m positions is in fact the remainder of $(x^m w(x) \pmod{x^n - 1})$ divided by $g(x)$. Put

$$r(x) := ((x^m w(x) \pmod{x^n - 1}) \pmod{g(x)}) = (x^m w(x) \pmod{g(x)}).$$

The weight of $r(x)$ is clearly the same as the weight of $e(x)$, which is at most $\lfloor (d-1)/2 \rfloor$. This shows the existence of m.

The word $t(x) := (x^{n-m} s_m(x) \pmod{x^n - 1})$ is a cyclic shift of $(s_m, \mathbf{0})$ through $n - m$ positions, where s_m is the vector of $\mathbf{F}_q^{n-k}$ corresponding to the polynomial $s_m(x)$. It is clear that the weight of $t(x)$ is the same as the weight of $s_m(x)$. Hence, $\mathrm{wt}(t(x)) \le \lfloor (d-1)/2 \rfloor$. As

$$x^m(w(x) - t(x)) \equiv x^m(w(x) - x^{n-m} s_m(x))$$
$$\equiv x^m w(x) - x^n s_m(x)$$
$$\equiv s_m(x) - x^n s_m(x)$$
$$\equiv (1 - x^n) s_m(x)$$
$$\equiv 0 \pmod{g(x)}$$

Table 7.3.

i	$s_i(x)$
0	$1 + x + x^2$
1	$1 + x$
2	$x + x^2$
3	1

Table 7.4.

i	$s_i(x)$
0	$1 + x^2 + x^5 + x^7$
1	$1 + x + x^3 + x^4 + x^7$
2	$1 + x + x^2 + x^5 + x^6 + x^7$
3	$1 + x + x^2 + x^3 + x^4$
4	$x + x^2 + x^3 + x^4 + x^5$
5	$x^2 + x^3 + x^4 + x^5 + x^6$
6	$x^3 + x^4 + x^5 + x^6 + x^7$
7	$1 + x^5$

and x^m is co-prime to $g(x)$ (see Remark 3.2.5(iii)), we claim that $w(x) - t(x)$ is divisible by $g(x)$; i.e., $w(x) - t(x)$ is a codeword. As $t(x)$ and the error pattern $e(x)$ are in the same coset, we have that $e(x) = t(x) = (x^{n-m}s_m(x) \pmod{x^n - 1})$ by Exercise 4.44.   □

Example 7.4.10 (i) As in Example 7.4.4, consider the received word

$$w_1(x) = 1011100 = 1 + x^2 + x^3 + x^4.$$

Compute the syndromes $s_i(x)$ of $x^i w_1(x)$ until $\mathrm{wt}(s_i(x)) \leq 1$ (Table 7.3). Decode $w_1(x) = 1011100$ to $w_1(x) - x^4 s_3(x) = w_1(x) - x^4 = 1 + x^2 + x^3 = 1011000$.

(ii) Consider the binary [15, 7]-cyclic code generated by $g(x) = 1 + x^4 + x^6 + x^7 + x^8$. We can check from the parity-check matrices that the minimum distance is 5. An error pattern with weight at most 2 must have a cyclic run of 0 of length at least 7. Thus, we can correct such an error pattern using the above algorithm. Consider the received word

$$w(x) = 110011101100010 = 1 + x + x^4 + x^5 + x^6 + x^8 + x^9 + x^{13}.$$

Compute the syndromes $s_i(x)$ of $x^i w(x)$ until $\mathrm{wt}(s_i(x)) \leq 2$ (Table 7.4).

Decode $w(x) = 110011101100010$ to $w(x) - x^8 s_7(x) = w(x) - x^8 - x^{13} = 1 + x + x^4 + x^5 + x^6 + x^9 = 110011100100000$.

7.5   Burst-error-correcting codes

So far, we have been concerned primarily with codes that correct random errors. However, there are certain communication channels, such as telephone lines and magnetic storage systems, which are affected by errors localized in short intervals rather than at random. Such an error is called a *burst error*. In general, codes for correcting random errors are not efficient for correcting burst errors. Therefore, it is desirable to construct codes specifically for correcting burst errors. Codes of this kind are called *burst-error-correcting codes*.

Cyclic codes are very efficient for correcting burst errors. Many effective cyclic burst-error-correcting codes have been found since the late 1970s. In this section, we will discuss some properties of burst-error-correcting codes and a decoding algorithm. The codes in this section are all binary codes.

Definition 7.5.1 A *burst* of length $l > 1$ is a binary vector whose nonzero components are confined to l cyclically consecutive positions, with the first and last positions being nonzero.

A code is called an *l-burst-error-correcting code* if it can correct all burst errors of length l or less; i.e., error patterns that are bursts of length l or less.

Example 7.5.2 0011010000 is a burst of length 4, while 01000000000000100 is a burst of length 5.

Theorem 7.5.3 *A linear code C is an l-burst-error-correcting code if and only if all the burst errors of length l or less lie in distinct cosets of C.*

Proof. If all the burst errors of length l or less lie in distinct cosets, then each burst error is determined by its syndrome. The error can then be corrected through its syndrome.

On the other hand, suppose that two distinct burst errors $\mathbf{b}_1$ and $\mathbf{b}_2$ of length l or less lie in the same coset of C. The difference $\mathbf{c} = \mathbf{b}_1 - \mathbf{b}_2$ is a codeword. Thus, if $\mathbf{b}_1$ is received, then $\mathbf{b}_1$ could be decoded to both $\mathbf{0}$ and $\mathbf{c}$. □

Corollary 7.5.4 *Let C be an $[n, k]$-linear l-burst-error-correcting code. Then*

(i) *no nonzero burst of length $2l$ or less can be a codeword;*
(ii) *(Reiger bound.) $n - k \geq 2l$.*

Proof. (i) Suppose that there exists a codeword $\mathbf{c}$ which is a burst of length $\leq 2l$. Then, $\mathbf{c}$ is of the form $(\mathbf{0}, 1, \mathbf{u}, \mathbf{v}, 1, \mathbf{0})$, where $\mathbf{u}$ and $\mathbf{v}$ are two words of length $\leq l - 1$. Hence, the words $\mathbf{w} = (\mathbf{0}, 1, \mathbf{u}, \mathbf{0}, 0, \mathbf{0})$ and $\mathbf{c} - \mathbf{w} = (\mathbf{0}, 0, \mathbf{0}, \mathbf{v}, 1, \mathbf{0})$ are two bursts of length $\leq l$. They are in the same coset. This is a contradiction to Theorem 7.5.3.

(ii) Let $\mathbf{u}_1, \mathbf{u}_2, \ldots, \mathbf{u}_{n-k+1}$ be the first $n - k + 1$ column vectors of a parity-check matrix of C. Then, they lie in $\mathbf{F}_2^{n-k}$ and are hence linearly dependent. Thus, there exist $c_1, c_2, \ldots, c_{n-k+1} \in \mathbf{F}_2$, not all zero, such that $\sum_{i=1}^{n-k+1} c_i \mathbf{u}_i = \mathbf{0}$. This implies that $(c_1, c_2, \ldots, c_{n-k+1}, \mathbf{0})$ is a codeword, and it is clear that this codeword is a burst of length $\leq n - k + 1$. By part (i), we have $n - k + 1 > 2l$; i.e., $n - k \geq 2l$. $\qquad\square$

An $[n, k]$-linear l-burst-error-correcting code satisfies $n - k \geq 2l$; i.e.,

$$l \leq \left\lfloor \frac{n - k}{2} \right\rfloor. \tag{7.7}$$

A linear burst-error-correcting code achieving the above Reiger bound is called an *optimal* burst-error-correcting code.

Example 7.5.5 Let C be the binary cyclic code of length 15 generated by $1 + x + x^2 + x^3 + x^6$. It is a $[15, 9]$-linear code. The reader may check that all the bursts of length 3 or less lie in distinct cosets of C. By Theorem 7.5.3, C is a 3-burst-error-correcting code. The reader may also want to confirm Corollary 7.5.4(i) by checking that no burst errors of length 6 or less are codewords. This code is optimal as the Reiger bound (7.7) is achieved.

Note that a burst of length l has a run of 0 of length $n - l$. By Corollary 7.5.4, we have $k \leq n - 2l \leq n - l$ for an $[n, k]$-linear l-burst-error-correcting code. This satisfies the requirement for the decoding algorithm in Section 7.4 to correct an error containing a cyclic run of at least k zeros. Hence, the algorithm can be directly employed to correct burst errors. The main difference is that, in the case of burst-error-correction, we do not require the weight of an error pattern to be less than or equal to $\lfloor (d(C) - 1)/2 \rfloor$. The modified decoding algorithm for burst-error-correction is as follows.

Decoding algorithm for cyclic burst-error-correcting codes

Let C be a q-ary $[n, k]$-cyclic code with generator polynomial $g(x)$. Let $w(x)$ be a received word with an error pattern $e(x)$ that is a burst error of length l or less.

Table 7.5.

i	$s_i(x)$
0	$1 + x + x^4 + x^5$
1	$1 + x^3 + x^5$
2	$1 + x^2 + x^3 + x^4$
3	$x + x^3 + x^4 + x^5$
4	$1 + x + x^3 + x^4 + x^5$
5	$1 + x^3 + x^4 + x^5$
6	$1 + x^2 + x^3 + x^4 + x^5$
7	$1 + x^2 + x^4 + x^5$
8	$1 + x^2 + x^5$
9	$1 + x^2$

Table 7.6.

Code parameters	Generator polynomials
$[7, 3]$	$1 + x + x^3 + x^4$
$[15, 9]$	$1 + x + x^2 + x^3 + x^6$
$[15, 7]$	$1 + x^4 + x^6 + x^7 + x^8$
$[15, 5]$	$1 + x + x^2 + x^4 + x^5 + x^8 + x^{10}$

Step 1: Compute the syndromes of $x^i w(x)$ for $i = 1, 2, \ldots$, and denote by $s_i(x)$ the syndrome of $x^i w(x)$.

Step 2: Find m such that the syndrome for $x^m w(x)$ is a burst of length l or less.

Step 3: Compute the remainder $e(x)$ of $x^{n-m} s_m(x)$ divided by $x^n - 1$. Decode $w(x)$ to $w(x) - e(x)$.

The proof of the algorithm is similar to the one in the previous section. Now we use the code in Example 7.5.5 to illustrate the above algorithm.

Example 7.5.6 Consider the binary $[15, 9]$-cyclic code generated by $g(x) = 1 + x + x^2 + x^3 + x^6$. We can correct all burst errors of length 3 or less. Suppose

$$w(x) = 111011101100000 = 1 + x + x^2 + x^4 + x^5 + x^6 + x^8 + x^9.$$

Compute the syndromes $s_i(x)$ of $x^i w(x)$ until $s_m(x)$ is a burst of length 3 or less (Table 7.5). Decode $w(x) = 111011101100000$ to $w(x) - x^6 s_9(x) = w(x) - x^6 - x^8 = 1 + x + x^2 + x^4 + x^5 + x^9 = 111011000100000$.

We end our discussion with a list of a few optimal burst-error-correcting cyclic codes (see Table 7.6).

Exercises

7.1 Which of the following codes are cyclic ones?
(a) $\{(0, 0, 0), (1, 1, 1), (2, 2, 2)\} \subset \mathbf{F}_3^3$;
(b) $\{(0, 0, 0), (1, 0, 0), (0, 1, 0), (0, 0, 1)\} \subset \mathbf{F}_q^3$;
(c) $\{(x_0, x_1, \ldots, x_{n-1}) \in \mathbf{F}_q^n : \sum_{i=0}^{n-1} x_i = 0\}$;
(d) $\{(x_0, x_1, \ldots, x_{n-1}) \in \mathbf{F}_8^n : \sum_{i=0}^{n-1} x_i^2 = 0\}$;
(e) $\{(x_0, x_1, \ldots, x_{n-1}) \in \mathbf{F}_2^n : \sum_{i=0}^{n-1} (x_i^2 + x_i) = 0\}$.

7.2 Show that the dual code of a cyclic code is cyclic.

7.3 Show that the set $I = \{f(x) \in \mathbf{F}_q[x] : f(0) = f(1) = 0\}$ is an ideal of $\mathbf{F}_q[x]$ and find a generator.

7.4 Suppose that x, y are two independent variables. Show that the polynomial ring $\mathbf{F}_q[x, y]$ is not a principal ideal ring.

7.5 Find all the possible monic generators for each of the following ideals:
(a) $I = < 1 + x + x^3 > \subset \mathbf{F}_2[x]/(x^7 - 1)$;
(b) $I = < 1 + x^2 > \subset \mathbf{F}_3[x]/(x^4 - 1)$.

7.6 Determine whether the following polynomials are generator polynomials of cyclic codes of given lengths:
(a) $g(x) = 1 + x + x^2 + x^3 + x^4$ for a binary cyclic code of length 7;
(b) $g(x) = 2 + 2x^2 + x^3$ for a ternary cyclic code of length 8;
(c) $g(x) = 2 + 2x + x^3$ for a ternary cyclic code of length 13.

7.7 For each of the following cyclic codes, find the corresponding generator polynomial:
(a) $\{\lambda(1, 1, \ldots, 1) : \lambda \in \mathbf{F}_q\} \subset \mathbf{F}_q^n$;
(b) $\{0000, 1010, 0101, 1111\} \subset \mathbf{F}_2^4$;
(c) $\{(x_0, x_1, \ldots, x_{n-1}) \in \mathbf{F}_q^n : \sum_{i=0}^{n-1} x_i = 0\}$;
(d) $\{(x_0, x_1, \ldots, x_{n-1}) \in \mathbf{F}_2^n : \sum_{i=0}^{n-1} x_i^3 = 0\}$.

7.8 Determine the smallest length for a binary cyclic code for which each of the following polynomials is the generator polynomial:
(a) $g(x) = 1 + x^4 + x^5$;
(b) $g(x) = 1 + x + x^2 + x^4 + x^6$.

7.9 Based on Example 3.4.13(ii), determine the following:
(a) the number of binary cyclic codes of length 21;
(b) all values k for which there exists a binary $[21, k]$-cyclic code;
(c) the number of binary $[21, 12]$-cyclic codes;
(d) the generator polynomial for each of the binary $[21, 12]$-cyclic codes.

7.10 Based on Example 3.4.13(i), determine the following:
  (a) the number of ternary cyclic codes of length 13;
  (b) all values k for which there exists a ternary $[13, k]$-cyclic code;
  (c) the number of ternary $[13, 7]$-cyclic codes;
  (d) the generator polynomial for each of the ternary $[13, 7]$-cyclic codes.

7.11 Construct the generator polynomials of all binary cyclic codes of length 15.

7.12 Let $g(x) = (1 + x)(1 + x + x^3) \in F_2[x]$ be the generator polynomial of a binary $[7, 3]$-cyclic code C. Write down a generator matrix and a parity-check matrix for C. Construct a generator matrix of the form $(I_3 | A)$.

7.13 Let $g(x) = 1 + x^4 + x^6 + x^7 + x^8 \in F_2[x]$ be the generator polynomial of a binary $[15, 7]$-cyclic code C. Write down a generator matrix and a parity-check matrix for C. Construct a generator matrix of the form $(I_7 | A)$.

7.14 Suppose a generator (or parity-check, respectively) matrix of a linear code C has the property that the cyclic shift of every row is still a codeword (or a codeword in the dual code, respectively). Show that C is a cyclic code.

7.15 Let $g_1(x), g_2(x)$ be the generator polynomials of the q-ary cyclic codes C_1, C_2 of the same length, respectively. Show that $C_1 \subseteq C_2$ if and only if $g_1(x)$ is divisible by $g_2(x)$.

7.16 Let $\mathbf{v} \in F_q^n$. Show that the generator polynomial of the smallest cyclic code containing $\mathbf{v}$ is equal to $\gcd(v(x), x^n - 1)$, where $v(x)$ is the polynomial corresponding to $\mathbf{v}$.

7.17 Determine the generator polynomial and the dimension of the smallest cyclic code containing each of the following words, respectively:
  (a) $1000111 \in F_2^7$;
  (b) $(1, 0, 2, 0, 2, 0, 1, 1) \in F_3^8$;
  (c) $101010111110010 \in F_2^{15}$.

7.18 Let $g(x)$ be the generator polynomial of a q-ary cyclic code C of length n. Put $h(x) = (x^n - 1)/g(x)$. Show that, if $a(x)$ is a polynomial satisfying $\gcd(a(x), h(x)) = 1$, then $a(x)g(x)$ is a generator of C. Conversely, if $g_1(x)$ is a generator of C, then there exists a polynomial $a(x)$ satisfying $\gcd(a(x), h(x)) = 1$ such that $g_1(x) \equiv a(x)g(x) \pmod{x^n - 1}$.

7.19 (a) Show that, for any $1 \le k \le 26$, there exists a ternary cyclic code of length 27 and dimension k.
  (b) Based on the factorization of $x^{15} - 1 \in F_2[x]$, show that, for any $1 \le k \le 15$, there exists a binary cyclic code of length 15 and dimension k.

7.20 Let α be a primitive element of $\mathbf{F}_{2^m}$ and let $g(x) \in \mathbf{F}_2[x]$ be the minimal polynomial of α with respect to $\mathbf{F}_2$. Show that the cyclic code of length $2^m - 1$ with $g(x)$ as the generator polynomial is in fact a binary $[2^m - 1, 2^m - 1 - m, 3]$-Hamming code.

7.21 Let C be a binary cyclic code of length $n \geq 3$ with generator polynomial $g(x) \neq 1$, where n is the smallest positive integer for which $x^n - 1$ is divisible by $g(x)$. Show that C has minimum distance at least 3. Is the result true for nonbinary cyclic codes?

7.22 Let $g(x)$ be the generator polynomial of a binary cyclic code C. Show that the subset C_E of even-weight vectors in C is also a cyclic code. Determine the generator polynomial of C_E in terms of $g(x)$.

7.23 Let C_i be a q-ary cyclic code of length n with generator polynomial $g_i(x)$, for $i = 1, 2$.
  (i) Show that $C_1 \cap C_2$ and $C_1 + C_2$ are both cyclic codes.
  (ii) Determine the generator polynomials of $C_1 \cap C_2$ and $C_1 + C_2$ in terms of $g_1(x)$, $g_2(x)$.

7.24 A codeword $e(x)$ of a q-ary cyclic code C of length n is called an *idempotent* if $e^2(x) \equiv e(x) \pmod{x^n - 1}$. If an idempotent $e(x)$ is also a generator of C, it is called a *generating idempotent*. Let $g(x)$ be the generator polynomial of a q-ary cyclic code C and put $h(x) = (x^n - 1)/g(x)$. Show that, if $\gcd(g(x), h(x)) = 1$, then C has a unique generating idempotent. In particular, show that, if $\gcd(n, q) = 1$, then there always exists a unique generating idempotent for a q-ary cyclic code of length n.

7.25 Find the generating idempotent for each of the following cyclic codes:
  (a) the binary $[7, 4]$-Hamming code Ham$(3, 2)$;
  (b) the binary $[15, 11, 3]$-Hamming code Ham$(4, 2)$;
  (c) the ternary $[13, 10, 3]$-Hamming code Ham$(3, 3)$.

7.26 Let C_i be a q-ary cyclic code of length n with generating idempotent $e_i(x)$ $(i = 1, 2)$. Show that $C_1 \cap C_2$ and $C_1 + C_2$ have generating idempotents $e_1(x)e_2(x)$ and $e_1(x) + e_2(x) - e_1(x)e_2(x)$, respectively.

7.27 An error pattern $\mathbf{e}$ of a code C is said to be *detectable* if $\mathbf{e} + \mathbf{c} \notin C$ for all $\mathbf{c} \in C$. Show that, for a cyclic code, if an error pattern $e(x)$ is detectable, then its ith cyclic shift is also detectable, for any i.

7.28 Let C be a binary $[7, 4]$-Hamming code with generator polynomial $g(x) = 1 + x + x^3$. Suppose each of the following received words has at most one error. Decode these words using error trapping:
  (a) 1101011;
  (b) 0101111;
  (c) 0100011.

7.29 A binary [15, 7]-cyclic code is generated by $g(x) = 1 + x^4 + x^6 + x^7 + x^8$. Decode the following received words using error trapping:
(a) 110111101110110;
(b) 111110100001000.

7.30 A binary [15, 5]-cyclic code is generated by $g(x) = 1 + x + x^2 + x^4 + x^5 + x^8 + x^{10}$. Construct a parity-check matrix of the form $(I_{10}|A)$. Decode the following words using error trapping:
(a) 011111110101000;
(b) 100101111011100.

7.31 Let C be the binary cyclic code of length 15 generated by $g(x) = 1 + x^2 + x^4 + x^5$.
(i) Find the minimum distance of C.
(ii) Show that C can correct all bursts of length 2 or less.
(iii) Decode the following received words using burst-error-correction:
(a) 010110000000010; (b) 110000111010011.

7.32 Let C be the binary [15, 9]-cyclic code generated by $g(x) = 1 + x^3 + x^4 + x^5 + x^6$. Decode the following received words using burst-error-correction:
(a) 101011101011100;
(b) 010000001011111.

7.33 Let α be a primitive element of $\mathbf{F}_{2^m}$ ($m > 2$) and let $g(x) \in \mathbf{F}_2[x]$ be its minimal polynomial with respect to $\mathbf{F}_2$. Let C be the binary cyclic code of length $2^m - 1$ generated by $(x + 1)g(x)$. An error pattern of the form

$$e(x) = x^i + x^{i+1}$$

is called a *double-adjacent-error pattern*. Show that no double-adjacent-error patterns can be in the same coset of C. Thus, C can correct all the single-error patterns and double-adjacent-error patterns.

7.34 Let $g_1(x)$, $g_2(x)$ be two polynomials over $\mathbf{F}_q$. Let n_i be the length of the shortest cyclic code that $g_i(x)$ generates, for $i = 1, 2$. Determine the length of the shortest cyclic code that $g_1(x)g_2(x)$ generates.

7.35 Let C be a binary cyclic code with generator polynomial $g(x)$.
(i) Prove that, if $g(x)$ is divisible by $x - 1$, then all the codewords have even weight.
(ii) Suppose the length of C is odd. Show that the all-one vector is a codeword if and only if $g(x)$ is not divisible by $x - 1$.
(iii) Suppose the length of C is odd. Show that C contains a codeword of odd weight if and only if the all-one vector is a codeword.

7.36 Let $g(x)$ be the generator polynomial of a q-ary $[n, k]$-cyclic code with $\gcd(n, q) = 1$. Show that the all-one vector is a codeword if and only if $g(x)$ is not divisible by $x - 1$.

7.37 Let C be a q-ary $[q + 1, 2]$-linear code with minimum distance q. Show that, if q is odd, then C is not a cyclic code.

7.38 Let n be a positive integer and $\gcd(n, q) = 1$. Assume that there are exactly t elements in a complete set of representatives of cyclotomic cosets of q modulo n. Show that there is a total of 2^t cyclic codes of length n over F_q.

7.39 Let $a \in F_q^*$. A q-ary linear code C is called *constacyclic* with respect to a if $(ac_{n-1}, c_0, \ldots, c_{n-2})$ belongs to C whenever $(c_0, c_1, \ldots, c_{n-1})$ belongs to C. In particular, C is called *negacyclic* if $a = -1$.

(a) Show that a q-ary linear code C of length n is constacyclic with respect to a if and only if $\pi_a(C)$ is an ideal of $F_q[x]/(x^n - a)$, where π_a is the map defined by

$$F_q^n \to F_q[x]/(x^n - a), \quad (c_0, c_1, \ldots, c_{n-1}) \mapsto \sum_{i=0}^{n-1} c_i x^i.$$

(b) Determine all the ternary negacyclic codes of length 8.

7.40 Suppose $x^n + 1$ has the factorization over F_q

$$\prod_{i=1}^{r} p_i^{e_i}(x),$$

where $e_i \geq 1$ and $p_i(x)$ are distinct monic irreducible polynomials. Find the number of q-ary negacyclic codes.

7.41 Let n be a positive integer with $\gcd(n, q) = 1$. Let α be a primitive nth root of unity in some extension field of F_q. Let $g(x) \in F_q[x]$ be the minimal polynomial of α with respect to F_q. Assume that the degree of $g(x)$ is m. Let C be the q-ary cyclic code with generator polynomial $g(x)$. Then the dual code $C^\perp$ is called an *irreducible cyclic code*. Show that $C^\perp$ is the trace code $\mathrm{Tr}_{F_{q^m}/F_q}(V)$, where V is the one-dimensional F_{q^m}-vector space $< (1, \alpha, \alpha^2, \ldots, \alpha^{n-1}) >$.

7.42 Let n be a positive integer and let $1 \leq \ell < n$ be a divisor of n. A linear code C over F_q is *quasi-cyclic of index ℓ* (or *ℓ-quasi-cyclic*) if

$$(c_{n-\ell}, c_{n-\ell+1}, \ldots, c_{n-1}, c_0, c_1, \ldots, c_{n-\ell-1}) \in C$$

whenever $(c_0, c_1, \ldots, c_{n-1}) \in C$. In particular, a 1-quasi-cyclic code is a cyclic code.

(a) Show that the dual of an ℓ-quasi-cyclic code is again ℓ-quasi-cyclic.

(b) For every positive integer m, show that there exist self-dual 2-quasi-cyclic codes over $\mathbf{F}_q$ of length $2m$ if q satisfies one of the following conditions:

    (i) q is a power of 2;

    (ii) $q = p^b$, where p is a prime congruent to 1 (mod 4);

    (iii) $q = p^{2b}$, where p is a prime congruent to 3 (mod 4).

7.43 Assume that $q \geq 3$ is a power of an odd prime. Let C_1, C_2 be two linear codes over $\mathbf{F}_q$ of length n.

    (i) Using notation as in Exercise 6.6, show that $C_1 \lozenge C_2$ is a quasi-cyclic code of index n.

    (ii) Show that every quasi-cyclic code over $\mathbf{F}_q$ of length $2n$ of index n is of the form $C_1 \lozenge C_2$ for some suitably chosen linear codes C_1 and C_2 over $\mathbf{F}_q$.

7.44 For a prime power $q \geq 2$, let $1 < m \leq q - 1$ be a divisor of $q - 1$ and let $\alpha \in \mathbf{F}_q^*$ be an element of order m. Let $C_0, C_1, \ldots, C_{m-1}$ be linear codes over $\mathbf{F}_q$ of length ℓ. Show that

$$C := \left\{ (\mathbf{x}_0, \mathbf{x}_1, \ldots, \mathbf{x}_{m-1}) : \mathbf{x}_i = \sum_{j=0}^{m-1} \alpha^{ij} \mathbf{c}_j, \right.$$
$$\left. \text{where } \mathbf{c}_j \in C_j \text{ for } 0 \leq j \leq m - 1 \right\}$$

is a quasi-cyclic code of length ℓm of index ℓ. (Note: the code C is called the *Vandermonde product* of $C_0, C_1, \ldots, C_{m-1}$.)

7.45 (a) Let q be an even prime power and let C_1, C_2 be linear codes over $\mathbf{F}_q$ of length n. Show that (see Exercise 6.7)

$$C := \{(\mathbf{a} + \mathbf{x}, \mathbf{b} + \mathbf{x}, \mathbf{a} + \mathbf{b} + \mathbf{x}) : \mathbf{a}, \mathbf{b} \in C_1, \mathbf{x} \in C_2\}$$

is an n-quasi-cyclic code over $\mathbf{F}_q$ of length $3n$.

(b) Let q be a power of an odd prime such that -1 is not a square in $\mathbf{F}_q$. Let i be an element of $\mathbf{F}_{q^2}$ such that $i^2 + 1 = 0$. Let Tr denote the trace $\mathrm{Tr}_{\mathbf{F}_{q^2}/\mathbf{F}_q}$ defined in Exercise 4.5. Let C_1, C_2 be linear codes over $\mathbf{F}_q$ of length ℓ and let C_3 be a linear code of length ℓ over $\mathbf{F}_{q^2}$. Show that

$$C := \left\{ (\mathbf{c}_0, \mathbf{c}_1, \mathbf{c}_2, \mathbf{c}_3) : \mathbf{c}_j = \mathbf{x} + (-1)^j \mathbf{y} \right.$$
$$\left. + \mathrm{Tr}(\mathbf{z} i^j), \mathbf{x} \in C_1, \mathbf{y} \in C_2, \mathbf{z} \in C_3 \right\}$$

is an ℓ-quasi-cyclic code over $\mathbf{F}_q$ of length 4ℓ.

8 Some special cyclic codes

The preceding chapter covered the subject of general cyclic codes. The structure of cyclic codes was analyzed, and two simple decoding algorithms were introduced. In particular, we showed that a cyclic code is totally determined by its generator polynomial. However, in general it is difficult to obtain information on the minimum distance of a cyclic code from its generator polynomial, even though the former is completely determined by the latter. On the other hand, if we choose some special generator polynomials properly, then information on the minimum distance can be gained, and also simpler decoding algorithms could apply. In this chapter, by carefully choosing the generator polynomials, we obtain several important classes of cyclic codes, such as BCH codes, Reed–Solomon codes and quadratic-residue codes. In addition to their structures, we also discuss a decoding algorithm for BCH codes.

8.1 BCH codes

The class of Bose, Chaudhuri and Hocquenghem (BCH) codes is, in fact, a generalization of the Hamming codes for multiple-error correction (recall that Hamming codes correct only one error). Binary BCH codes were first discovered by A. Hocquenghem [8] in 1959 and independently by R. C. Bose and D. K. Ray-Chaudhuri [1] in 1960. Generalizations of the binary BCH codes to q-ary codes were obtained by D. Gorenstein and N. Zierler [5] in 1961.

8.1.1 Definitions

We defined the least common multiple $\text{lcm}(f_1(x), f_2(x))$ of two nonzero polynomials $f_1(x)$, $f_2(x) \in \mathbf{F}_q[x]$ to be the monic polynomial of the lowest degree which is a multiple of both $f_1(x)$ and $f_2(x)$ (see Chapter 3). Suppose we have t

nonzero polynomials $f_1(x), f_2(x), \ldots, f_t(x) \in \mathbf{F}_q[x]$. The *least common multiple* of $f_1(x), \ldots, f_t(x)$ is the monic polynomial of the lowest degree which is a multiple of all of $f_1(x), \ldots, f_t(x)$, denoted by $\mathrm{lcm}(f_1(x), \ldots, f_t(x))$.

It can be proved that the least common multiple of the polynomials $f_1(x)$, $f_2(x)$, $f_3(x)$ is the same as $\mathrm{lcm}(\mathrm{lcm}(f_1(x), f_2(x)), f_3(x))$ (see Exercise 8.2).

By induction, one can prove that the least common multiple of the polynomials $f_1(x)$, $f_2(x),\ldots, f_t(x)$ is the same as $\mathrm{lcm}(\mathrm{lcm}(f_1(x), \ldots, f_{t-1}(x)), f_t(x))$.

Remark 8.1.1 If $f_1(x), \ldots, f_t(x) \in \mathbf{F}_q[x]$ have the following factorizations:

$$f_1(x) = a_1 \cdot p_1(x)^{e_{1,1}} \cdots p_n(x)^{e_{1,n}}, \ldots, f_t(x) = a_t \cdot p_1(x)^{e_{t,1}} \cdots p_n(x)^{e_{t,n}},$$

where $a_1, \ldots, a_t \in \mathbf{F}_q^*$, $e_{i,j} \geq 0$ and $p_i(x)$ are distinct monic irreducible polynomials over $\mathbf{F}_q$, then

$$\mathrm{lcm}(f_1(x), \ldots, f_t(x)) = p_1(x)^{\max\{e_{1,1}, \ldots, e_{t,1}\}} \cdots p_n(x)^{\max\{e_{1,n}, \ldots, e_{t,n}\}}.$$

Example 8.1.2 Consider the binary polynomials $f_1(x) = (1+x)^2(1+x+x^4)^3$, $f_2(x) = (1+x)(1+x+x^2)^2$, $f_3(x) = x^2(1+x+x^4)$. Then we have, by the above remark, that

$$\mathrm{lcm}(f_1(x), f_2(x), f_3(x)) = x^2(1+x)^2(1+x+x^2)^2(1+x+x^4)^3.$$

Lemma 8.1.3 *Let* $f(x), f_1(x), f_2(x), \ldots, f_t(x)$ *be polynomials over* $\mathbf{F}_q$. *If* $f(x)$ *is divisible by every polynomial* $f_i(x)$ *for* $i = 1, 2, \ldots, t$, *then* $f(x)$ *is divisible by* $\mathrm{lcm}(f_1(x), f_2(x), \ldots, f_t(x))$ *as well.*

Proof. Put $g(x) = \mathrm{lcm}(f_1(x), f_2(x), \ldots, f_t(x))$. By the division algorithm, there exist two polynomials $u(x)$ and $r(x)$ over $\mathbf{F}_q$ such that $\deg(r(x)) < \deg(g(x))$ and

$$f(x) = u(x)g(x) + r(x).$$

Thus, $r(x) = f(x) - u(x)g(x)$, and therefore $r(x)$ is also divisible by all $f_i(x)$. Since $g(x)$ has the smallest degree, this forces $r(x) = 0$. $\qquad\square$

Example 8.1.4 The polynomial $f(x) = x^{15} - 1 \in \mathbf{F}_2[x]$ is divisible by $f_1(x) = 1 + x + x^2 \in \mathbf{F}_2[x]$, $f_2(x) = 1 + x + x^4 \in \mathbf{F}_2[x]$, and $f_3(x) = (1 + x + x^2)(1 + x^3 + x^4) \in \mathbf{F}_2[x]$, respectively. Then $f(x)$ is also divisible by $\mathrm{lcm}(f_1(x), f_2(x), f_3(x)) = (1 + x + x^2)(1 + x + x^4)(1 + x^3 + x^4)$.

Example 8.1.5 Fix a primitive element α of $\mathbf{F}_{q^m}$ and denote by $M^{(i)}(x)$ the minimal polynomial of α^i with respect to $\mathbf{F}_q$. By Theorem 3.4.8, each root

β of $M^{(i)}(x)$ is an element of $\mathbf{F}_{q^m}$, and therefore β satisfies $\beta^{q^m-1} - 1 = 0$; i.e., $x - \beta$ is a linear divisor of $x^{q^m-1} - 1$. By Theorem 3.4.8 again, $M^{(i)}(x)$ has no multiple roots. Hence, $M^{(i)}(x)$ is a divisor of $x^{q^m-1} - 1$. For a subset I of $\mathbf{Z}_{q^m-1}$, the least common multiple $\mathrm{lcm}(M^{(i)}(x))_{i \in I}$ is a divisor of $x^{q^m-1} - 1$ as well by Lemma 8.1.3.

The above example provides a method to find some divisors of $x^{q^m-1} - 1$. These divisors can be chosen as generator polynomials of cyclic codes of length $q^m - 1$.

Definition 8.1.6 Let α be a primitive element of $\mathbf{F}_{q^m}$ and denote by $M^{(i)}(x)$ the minimal polynomial of α^i with respect to $\mathbf{F}_q$. A (*primitive*) *BCH code* over $\mathbf{F}_q$ of length $n = q^m - 1$ with designed distance δ is a q-ary cyclic code generated by $g(x) := \mathrm{lcm}(M^{(a)}(x), M^{(a+1)}(x), \ldots, M^{(a+\delta-2)}(x))$ for some integer a. Furthermore, the code is called *narrow-sense* if $a = 1$.

Example 8.1.7 (i) Let α be a primitive element of $\mathbf{F}_{2^m}$. Then a narrow-sense binary BCH code with designed distance 2 is a cyclic code generated by $M^{(1)}(x)$. It is in fact a Hamming code (see Exercise 7.20).

(ii) Let $\alpha \in \mathbf{F}_8$ be a root of $1 + x + x^3$. Then it is a primitive element of $\mathbf{F}_8$. The polynomials $M^{(1)}(x)$ and $M^{(2)}(x)$ are both equal to $1 + x + x^3$. Hence, a narrow-sense binary BCH code of length 7 generated by $\mathrm{lcm}(M^{(1)}(x), M^{(2)}(x)) = 1 + x + x^3$ is a [7, 4]-code. In fact, it is a binary [7, 4, 3]-Hamming code (see Exercise 7.20).

(iii) With α as in (ii), a binary BCH code of length 7 generated by $\mathrm{lcm}(M^{(0)}(x), M^{(1)}(x), M^{(2)}(x)) = \mathrm{lcm}(1 + x, 1 + x + x^3) = (1 + x)(1 + x + x^3)$ is a [7, 3]-cyclic code. It is easy to verify that this code is the dual code of the Hamming code of (ii).

Example 8.1.8 Let β be a root of $1 + x + x^2 \in \mathbf{F}_2[x]$, then $\mathbf{F}_4 = \mathbf{F}_2[\beta]$. Let α be a root of $\beta + x + x^2 \in \mathbf{F}_4[x]$. Then α is a primitive element of $\mathbf{F}_{16}$. Consider the narrow-sense 4-ary BCH code of length 15 with designed distance 4. Then the generator polynomial is

$$g(x) = \mathrm{lcm}(M^{(1)}(x), M^{(2)}(x), M^{(3)}(x)) = 1 + \beta x + \beta x^2 + x^3 + x^4 + \beta^2 x^5 + x^6.$$

8.1.2 Parameters of BCH codes

The length of a BCH code is clearly $q^m - 1$. We consider the dimension of BCH codes first.

Theorem 8.1.9 (i) *The dimension of a q-ary BCH code of length $q^m - 1$ generated by $g(x) := \text{lcm}(M^{(a)}(x), M^{(a+1)}(x), \ldots, M^{(a+\delta-2)}(x))$ is independent of the choice of the primitive element α.*

(ii) *A q-ary BCH code of length $q^m - 1$ with designed distance δ has dimension at least $q^m - 1 - m(\delta - 1)$.*

Proof. (i) Let C_i be the cyclotomic coset of q modulo $q^m - 1$ containing i. Put $S = \bigcup_{i=a}^{a+\delta-2} C_i$. By Theorem 3.4.8 and Remark 8.1.1, we have

$$g(x) = \text{lcm}\left(\prod_{i \in C_a}(x - \alpha^i), \prod_{i \in C_{a+1}}(x - \alpha^i), \ldots, \prod_{i \in C_{a+\delta-2}}(x - \alpha^i)\right) = \prod_{i \in S}(x - \alpha^i).$$

Hence, the dimension is equal to $q^m - 1 - \deg(g(x)) = q^m - 1 - |S|$. As the set S is independent of the choice of α, the desired result follows.

(ii) By part (i), the dimension k satisfies

$$k = q^m - 1 - |S|$$
$$= q^m - 1 - \left|\bigcup_{i=a}^{a+\delta-2} C_i\right|$$
$$\geq q^m - 1 - \sum_{i=a}^{a+\delta-2} |C_i|$$
$$\geq q^m - 1 - \sum_{i=a}^{a+\delta-2} m \qquad \text{(by Remark 3.4.6(i))}$$
$$= q^m - 1 - m(\delta - 1).$$

This completes the proof. $\square$

The above result shows that, in order to find the dimension of a q-ary BCH code of length $q^m - 1$ generated by $g(x) := \text{lcm}(M^{(a)}(x), M^{(a+1)}(x), \ldots, M^{(a+\delta-2)}(x))$, it is sufficient to check the cardinality of $\bigcup_{i=a}^{a+\delta-2} C_i$, where C_i is the cyclotomic coset of q modulo $q^m - 1$ containing i.

Example 8.1.10 (i) Consider the following cyclotomic cosets of 2 modulo 15:

$$C_2 = \{1, 2, 4, 8\}, \quad C_3 = \{3, 6, 12, 9\}.$$

Then the dimension of the binary BCH code of length 15 of designed distance 3 generated by $g(x) := \text{lcm}(M^{(2)}(x), M^{(3)}(x))$ is

$$15 - |C_2 \cup C_3| = 15 - 8 = 7.$$

Note that the lower bound in Theorem 8.1.9(ii) is attained for this example.

Table 8.1.

n	k	t	n	k	t
7	4	1	63	51	2
15	11	1	63	45	3
15	7	2	63	39	4
15	5	3	63	36	5
31	26	1	63	30	6
31	21	2	63	24	7
31	16	3	63	18	10
31	11	5	63	16	11
31	6	7	63	10	13
63	57	1	63	7	15

(ii) Consider the following cyclotomic cosets of 3 modulo 26:

$$C_1 = C_3 = \{1, 3, 9\}, \quad C_2 = \{2, 6, 18\}, \quad C_4 = \{4, 12, 10\}.$$

Then the dimension of the ternary BCH code of length 26 and designed distance 5 generated by

$$g(x) := \mathrm{lcm}(M^{(1)}(x), M^{(2)}(x), M^{(3)}(x), M^{(4)}(x))$$

is

$$26 - |C_1 \cup C_2 \cup C_3 \cup C_4| = 26 - 9 = 17.$$

Note that, for this example, the dimension is strictly bigger than the lower bound in Theorem 8.1.9(ii).

Example 8.1.11 (i) For $t \geq 1$, t and $2t$ belong to the same cyclotomic coset of 2 modulo $2^m - 1$. This is equivalent to the fact that $M^{(t)}(x) = M^{(2t)}(x)$. Therefore,

$$\mathrm{lcm}(M^{(1)}(x), \ldots, M^{(2t-1)}(x)) = \mathrm{lcm}(M^{(1)}(x), \ldots, M^{(2t)}(x));$$

i.e., the narrow-sense binary BCH codes of length $2^m - 1$ with designed distance $2t + 1$ are the same as the narrow-sense binary BCH codes of length $2^m - 1$ with designed distance $2t$.

In Table 8.1 we list the dimensions of narrow-sense binary BCH codes of length $2^m - 1$ with designed distance $2t + 1$, for $3 \leq m \leq 6$. Note that the dimension of a narrow-sense BCH code is independent of the choice of the primitive elements (see Theorem 8.1.9(i)).

(ii) Let α be a root of $1 + x + x^4 \in \mathbf{F}_2[x]$. Then α is a primitive element of $\mathbf{F}_{16}$. By Example 3.4.7(i) and Theorem 3.4.8, we can compute the minimal

Table 8.2.

n	k	t	Generator polynomial
15	11	1	$1 + x + x^4$
15	7	2	$(1 + x + x^4)(1 + x + x^2 + x^3 + x^4)$
15	5	3	$(1 + x + x^4)(1 + x + x^2 + x^3 + x^4)(1 + x + x^2)$

polynomials

$$M^{(0)}(x) = 1 + x,$$
$$M^{(1)}(x) = M^{(2)}(x) = M^{(4)}(x) = M^{(8)}(x) = 1 + x + x^4,$$
$$M^{(3)}(x) = M^{(6)}(x) = M^{(12)}(x) = M^{(9)}(x) = 1 + x + x^2 + x^3 + x^4,$$
$$M^{(5)}(x) = M^{(10)}(x) = 1 + x + x^2,$$
$$M^{(7)}(x) = M^{(14)}(x) = M^{(13)}(x) = M^{(11)}(x) = 1 + x^3 + x^4.$$

The generator polynomials of the narrow-sense binary BCH codes of length 15 in Table 8.1 are given in Table 8.2.

Example 8.1.10(ii) shows that the lower bound in Theorem 8.1.9(ii) can be improved in some cases. The following result gives a sufficient condition under which the lower bound in Theorem 8.1.9(ii) can be achieved.

Proposition 8.1.12 *A narrow-sense q-ary BCH code of length $q^m - 1$ with designed distance δ has dimension exactly $q^m - 1 - m(\delta - 1)$ if $q \neq 2$ and $\gcd(q^m - 1, e) = 1$ for all $1 \leq e \leq \delta - 1$.*

Proof. From the proof of Theorem 8.1.9, we know that the dimension is equal to

$$q^m - 1 - \left| \bigcup_{i=1}^{\delta-1} C_i \right|,$$

where C_i stands for the cyclotomic coset of q modulo $q^m - 1$ containing i. Hence, it is sufficient to prove that $|C_i| = m$ for all $1 \leq i \leq \delta - 1$, and that C_i and C_j are disjoint for all $1 \leq i < j \leq \delta - 1$.

For any integer $1 \leq t \leq m - 1$, we claim that $i \not\equiv q^t i \pmod{q^m - 1}$ for $1 \leq i \leq \delta - 1$. Otherwise, we would have $(q^t - 1)i \equiv 0 \pmod{q^m - 1}$. This forces $(q^t - 1) \equiv 0 \pmod{q^m - 1}$ as $\gcd(i, q^m - 1) = 1$. This is a contradiction. This implies that $|C_i| = m$ for all $1 \leq i \leq \delta - 1$.

For any integers $1 \leq i < j \leq \delta - 1$, we claim that $j \not\equiv q^s i \pmod{q^m - 1}$ for any integer $s \geq 0$. Otherwise, we would have $j - i \equiv (q^s - 1)i \pmod{q^m - 1}$.

This forces $j - i \equiv 0 \pmod{q - 1}$, which is a contradiction to the condition $\gcd(j - i, q^m - 1) = 1$. Hence, C_i and C_j are disjoint. $\square$

Example 8.1.13 Consider a narrow-sense 4-ary BCH code of length 63 with designed distance 3. Its dimension is $63 - 3(3 - 1) = 57$.

As we know that a narrow-sense binary BCH code with designed distance $2t$ is the same as a narrow-sense binary BCH code with designed distance $2t + 1$, it is enough to consider narrow-sense binary BCH codes with odd designed distance.

Proposition 8.1.14 *A narrow-sense binary BCH code of length $n = 2^m - 1$ and designed distance $\delta = 2t + 1$ has dimension at least $n - m(\delta - 1)/2$.*

Proof. As the cyclotomic cosets C_{2i} and C_i are the same, the dimension k satisfies

$$k = 2^m - 1 - \left| \bigcup_{i=1}^{2t} C_i \right|$$

$$= 2^m - 1 - \left| \bigcup_{i=1}^{t} C_{2i-1} \right|$$

$$\geq 2^m - 1 - \sum_{i=1}^{t} |C_{2i-1}|$$

$$\geq 2^m - 1 - tm$$

$$= 2^m - 1 - m(\delta - 1)/2.$$

$\square$

Example 8.1.15 A narrow-sense binary BCH code of length 63 with designed distance $\delta = 5$ has dimension exactly $51 = 63 - 6(5 - 1)/2$. However, a narrow-sense binary BCH code of length 31 with designed distance $\delta = 11$ has dimension 11, which is bigger than $31 - 5(11 - 1)/2$.

For the rest of this subsection, we study the minimum distance of BCH codes.

Lemma 8.1.16 *Let C be a q-ary cyclic code of length n with generator polynomial $g(x)$. Suppose $\alpha_1, \ldots, \alpha_r$ are all the roots of $g(x)$ and the polynomial $g(x)$ has no multiple roots. Then an element $c(x)$ of $\mathbf{F}_q[x]/(x^n - 1)$ is a codeword of C if and only if $c(\alpha_i) = 0$ for all $i = 1, \ldots, r$.*

Proof. If $c(x)$ is a codeword of C, then there exists a polynomial $f(x)$ such that $c(x) = g(x)f(x)$. Thus, we have $c(\alpha_i) = g(\alpha_i)f(\alpha_i) = 0$ for all $i = 1, \dots, r$.

Conversely, if $c(\alpha_i) = 0$ for all $i = 1, \dots, r$, then $c(x)$ is divisible by $g(x)$ since $g(x)$ has no multiple roots. This means that $c(x)$ is a codeword of C. $\qquad\square$

Example 8.1.17 Consider the binary $[7, 4]$-Hamming code with generator polynomial $g(x) = 1 + x + x^3$. As all the elements of $\mathbf{F}_8\backslash\{0, 1\}$ are roots of $c(x) = 1 + x + x^2 + x^3 + x^4 + x^5 + x^6 = (x^7 - 1)/(x - 1)$, all the roots of $g(x)$ are roots of $c(x)$ as well. Thus, 1111111 is a codeword.

The following theorem explains the term 'designed distance'.

Theorem 8.1.18 *A BCH code with designed distance δ has minimum distance at least δ.*

Proof. Let α be a primitive element of $\mathbf{F}_{q^m}$ and let C be a BCH code generated by $g(x) := \mathrm{lcm}(M^{(a)}(x), M^{(a+1)}(x), \dots, M^{(a+\delta-2)}(x))$. It is clear that the elements $\alpha^a, \dots, \alpha^{a+\delta-2}$ are roots of $g(x)$.

Suppose that the minimum distance d of C is less than δ. Then there exists a nonzero codeword $c(x) = c_0 + c_1 x + \cdots + c_{n-1}x^{n-1}$ such that $\mathrm{wt}(c(x)) = d < \delta$. By Lemma 8.1.16, we have $c(\alpha^i) = 0$ for all $i = a, \dots, a + \delta - 2$; i.e.,

$$
\begin{pmatrix}
1 & \alpha^a & (\alpha^a)^2 & \cdots & (\alpha^a)^{n-1} \\
1 & \alpha^{a+1} & (\alpha^{a+1})^2 & \cdots & (\alpha^{a+1})^{n-1} \\
1 & \alpha^{a+2} & (\alpha^{a+2})^2 & \cdots & (\alpha^{a+2})^{n-1} \\
\vdots & \vdots & \vdots & \cdots & \vdots \\
1 & \alpha^{a+\delta-2} & (\alpha^{a+\delta-2})^2 & \cdots & (\alpha^{a+\delta-2})^{n-1}
\end{pmatrix}
\begin{pmatrix}
c_0 \\ c_1 \\ c_2 \\ \vdots \\ c_{n-1}
\end{pmatrix}
= \mathbf{0}. \qquad (8.1)
$$

Assume that the support of $c(x)$ is $R = \{i_1, \dots, i_d\}$, i.e., $c_j \neq 0$ if and only if $j \in R$. Then (8.1) becomes

$$
\begin{pmatrix}
(\alpha^a)^{i_1} & (\alpha^a)^{i_2} & (\alpha^a)^{i_3} & \cdots & (\alpha^a)^{i_d} \\
(\alpha^{a+1})^{i_1} & (\alpha^{a+1})^{i_2} & (\alpha^{a+1})^{i_3} & \cdots & (\alpha^{a+1})^{i_d} \\
(\alpha^{a+2})^{i_1} & (\alpha^{a+2})^{i_2} & (\alpha^{a+2})^{i_3} & \cdots & (\alpha^{a+2})^{i_d} \\
\vdots & \vdots & \vdots & \cdots & \vdots \\
(\alpha^{a+\delta-2})^{i_1} & (\alpha^{a+\delta-2})^{i_2} & (\alpha^{a+\delta-2})^{i_3} & \cdots & (\alpha^{a+\delta-2})^{i_d}
\end{pmatrix}
\begin{pmatrix}
c_{i_1} \\ c_{i_2} \\ c_{i_3} \\ \vdots \\ c_{i_d}
\end{pmatrix}
= \mathbf{0}. \quad (8.2)
$$

Since $d \leq \delta - 1$, we obtain the following system of equations by choosing the first d equations of the above system of equations:

$$
\begin{pmatrix}
(\alpha^a)^{i_1} & (\alpha^a)^{i_2} & (\alpha^a)^{i_3} & \cdots & (\alpha^a)^{i_d} \\
(\alpha^{a+1})^{i_1} & (\alpha^{a+1})^{i_2} & (\alpha^{a+1})^{i_3} & \cdots & (\alpha^{a+1})^{i_d} \\
(\alpha^{a+2})^{i_1} & (\alpha^{a+2})^{i_2} & (\alpha^{a+2})^{i_3} & \cdots & (\alpha^{a+2})^{i_d} \\
\cdot & \cdot & \cdot & & \cdot \\
\cdot & \cdot & \cdot & \cdots & \cdot \\
\cdot & \cdot & \cdot & & \cdot \\
(\alpha^{a+d-1})^{i_1} & (\alpha^{a+d-1})^{i_2} & (\alpha^{a+d-1})^{i_3} & \cdots & (\alpha^{a+d-1})^{i_d}
\end{pmatrix}
\begin{pmatrix}
c_{i_1} \\ c_{i_2} \\ c_{i_3} \\ \cdot \\ \cdot \\ \cdot \\ c_{i_d}
\end{pmatrix} = 0. \quad (8.3)
$$

The determinant D of the coefficient matrix of the above equation is equal to

$$
D = \prod_{j=1}^{d} (\alpha^a)^{i_j} \det
\begin{pmatrix}
1 & 1 & 1 & \cdots & 1 \\
\alpha^{i_1} & \alpha^{i_2} & \alpha^{i_3} & \cdots & \alpha^{i_d} \\
(\alpha^2)^{i_1} & (\alpha^2)^{i_2} & (\alpha^2)^{i_3} & \cdots & (\alpha^2)^{i_d} \\
\cdot & \cdot & \cdot & & \cdot \\
\cdot & \cdot & \cdot & \cdots & \cdot \\
\cdot & \cdot & \cdot & & \cdot \\
(\alpha^{d-1})^{i_1} & (\alpha^{d-1})^{i_2} & (\alpha^{d-1})^{i_3} & \cdots & (\alpha^{d-1})^{i_d}
\end{pmatrix} \quad (8.4)
$$

$$
= \prod_{j=1}^{d} (\alpha^a)^{i_j} \prod_{k>l} (\alpha^{i_k} - \alpha^{i_l}) \neq 0.
$$

Combining (8.3) and (8.4), we obtain $(c_{i_1}, \ldots, c_{i_d}) = 0$. This is a contradiction. $\qquad\square$

Example 8.1.19 (i) Let α be a root of $1 + x + x^3 \in \mathbf{F}_2[x]$, and let C be the binary BCH code of length 7 with designed distance 4 generated by

$$
g(x) = \mathrm{lcm}(M^{(0)}(x), M^{(1)}(x), M^{(2)}(x)) = 1 + x^2 + x^3 + x^4.
$$

Then $d(C) \leq \mathrm{wt}(g(x)) = 4$. On the other hand, we have, by Theorem 8.1.18, that $d(C) \geq 4$. Hence, $d(C) = 4$.

(ii) Let α be a root of $1 + x + x^4 \in \mathbf{F}_2[x]$. Then α is a primitive element of $\mathbf{F}_{16}$. Consider the narrow-sense binary BCH code of length 15 with designed distance 7. Then the generator polynomial is

$$
\begin{aligned}
g(x) &= \mathrm{lcm}(M^{(1)}(x), M^{(2)}(x), \ldots, M^{(6)}(x)) \\
&= M^{(1)}(x)M^{(3)}(x)M^{(5)}(x) \\
&= 1 + x + x^2 + x^4 + x^5 + x^8 + x^{10}.
\end{aligned}
$$

Therefore, $d(C) \leq \mathrm{wt}(g(x)) = 7$. On the other hand, we have, by Theorem 8.1.18, that $d(C) \geq 7$. Hence, $d(C) = 7$.

Example 8.1.20 Let α be a primitive element of $\mathbf{F}_{2^m}$ and let $M^{(1)}(x)$ be the minimal polynomial of α with respect to $\mathbf{F}_2$. Consider the narrow-sense binary BCH code C of length $n = 2^m - 1$ with designed distance 3 generated by

$$g(x) = \mathrm{lcm}(M^{(1)}(x), M^{(2)}(x)) = M^{(1)}(x).$$

Then, $d(C) \geq 3$ by Theorem 8.1.18. C is in fact a binary Hamming code by Exercise 7.20. Hence, $d(C) = 3$.

8.1.3   Decoding of BCH codes

The decoding algorithm we describe in this section is divided into three steps: (i) calculating the syndromes; (ii) finding the *error locator polynomial*; (iii) finding all roots of the error locator polynomial. For simplicity, we will discuss only the decoding of narrow-sense binary BCH codes.

Let C be a narrow-sense binary BCH code of length $n = 2^m - 1$ with designed distance $\delta = 2t + 1$ generated by $g(x) := \mathrm{lcm}(M^{(1)}(x), M^{(2)}(x)$ $, \ldots, M^{(\delta-1)}(x))$, where $M^{(i)}(x)$ is the minimal polynomial of α^i with respect to $\mathbf{F}_2$ for a primitive element α of $\mathbf{F}_{2^m}$.

Put

$$H = \begin{pmatrix} 1 & \alpha & (\alpha)^2 & \cdots & (\alpha)^{n-1} \\ 1 & \alpha^2 & (\alpha^2)^2 & \cdots & (\alpha^2)^{n-1} \\ 1 & \alpha^3 & (\alpha^3)^2 & \cdots & (\alpha^3)^{n-1} \\ \cdot & \cdot & \cdot & & \cdot \\ \cdot & \cdot & \cdot & \cdots & \cdot \\ \cdot & \cdot & \cdot & & \cdot \\ 1 & \alpha^{\delta-1} & (\alpha^{\delta-1})^2 & \cdots & (\alpha^{\delta-1})^{n-1} \end{pmatrix} \tag{8.5}$$

Then it can be shown that a word $\mathbf{c} \in \mathbf{F}_2^n$ is a codeword of C if and only if $\mathbf{c}H^{\mathrm{T}} = \mathbf{0}$ (see Exercise 8.9). Therefore, we can define the syndrome $S_{\mathrm{H}}(\mathbf{w})$ of a word $\mathbf{w} \in \mathbf{F}_2^n$ with respect to H by $\mathbf{w}H^{\mathrm{T}}$. Some properties of $S_{\mathrm{H}}(\mathbf{w})$ are also contained in Exercise 8.9.

Suppose that $w(x) = w_0 + w_1 x + \cdots + w_{n-1}x^{n-1}$ is a received word with the error polynomial $e(x)$ satisfying $\mathrm{wt}(e(x)) \leq t$. Put $c(x) = w(x) - e(x)$, then $c(x)$ is a codeword.

Step 1: Calculation of syndromes The syndrome of $w(x)$ is

$$(s_0, s_1, \ldots, s_{\delta-2}) := (w_0, w_1, \ldots, w_{n-1})H^{\mathrm{T}}.$$

It is clear that $s_i = w(\alpha^{i+1}) = e(\alpha^{i+1})$ for all $i = 0, 1, \ldots, \delta - 2$, since α^{i+1} are roots of $g(x)$.

Assume that the errors take place at positions $i_0, i_1, \ldots, i_{l-1}$ with $l \leq t$; i.e.,

$$e(x) = x^{i_0} + x^{i_1} + \cdots + x^{i_{l-1}}. \tag{8.6}$$

Then we obtain a system of equations

$$
\begin{aligned}
\alpha^{i_0} + \alpha^{i_1} + \cdots + \alpha^{i_{l-1}} &= s_0 = w(\alpha), \\
(\alpha^{i_0})^2 + (\alpha^{i_1})^2 + \cdots + (\alpha^{i_{l-1}})^2 &= s_1 = w(\alpha^2), \\
&\vdots \\
(\alpha^{i_0})^{\delta-1} + (\alpha^{i_1})^{\delta-1} + \cdots + (\alpha^{i_{l-1}})^{\delta-1} &= s_{\delta-2} = w(\alpha^{\delta-1}).
\end{aligned}
\tag{8.7}
$$

Any method for solving the above system of equations is a decoding algorithm for BCH codes.

Step 2: Finding the error locator polynomial For $e(x) = x^{i_0} + x^{i_1} + \cdots + x^{i_{l-1}}$, define the *error locator polynomial* by

$$\sigma(z) := \prod_{j=0}^{l-1} (1 - \alpha^{i_j} z).$$

It is clear that the error positions i_j can be found as long as all the roots of $\sigma(z)$ are known. For this step, we have to determine the error locator polynomial $\sigma(z)$.

Theorem 8.1.21 *Suppose the syndrome polynomial* $s(z) = \sum_{j=0}^{\delta-2} s_j z^j$ *is not the zero polynomial. Then there exists a nonzero polynomial* $r(z) \in \mathbf{F}_{2^m}[z]$ *such that* $\deg(r(z)) \leq t - 1$, $\gcd(r(z), \sigma(z)) = 1$ *and*

$$r(z) \equiv s(z)\sigma(z) \,(\mathrm{mod}\ z^{\delta-1}). \tag{8.8}$$

Moreover, for any pair $(u(z), v(z))$ *of nonzero polynomials over* $\mathbf{F}_{2^m}$ *satisfying* $\deg(u(z)) \leq t - 1$, $\deg(v(z)) \leq t$ *and*

$$u(z) \equiv s(z)v(z) \,(\mathrm{mod}\ z^{\delta-1}), \tag{8.9}$$

we have

$$\sigma(z) = \beta v(z), \qquad r(z) = \beta u(z), \tag{8.10}$$

for a nonzero element $\beta \in \mathbf{F}_{2^m}$.

Proof. (Uniqueness.) Multiplying (8.8) by $v(z)$ and (8.9) by $\sigma(z)$ gives

$$v(z)r(z) \equiv \sigma(z)u(z) \,(\mathrm{mod}\ z^{\delta-1}). \tag{8.11}$$

As $\deg(v(z)r(z)) \le 2t - 1 = \delta - 2$ and $\deg(\sigma(z)u(z)) \le 2t - 1 = \delta - 2$, it follows from (8.11) that $v(z)r(z) = \sigma(z)u(z)$. By the conditions that $\gcd(r(z), \sigma(z)) = 1$ and all the polynomials are nonzero, we obtain $\sigma(z) = \beta v(z)$ and $r(z) = \beta u(z)$ for a nonzero element $\beta \in \mathbf{F}_{2^m}$.

(Existence.) Put

$$r(z) = \sigma(z) \sum_{j=0}^{l-1} \frac{\alpha^{i_j}}{(1 - \alpha^{i_j}z)}.$$

Then

$$\frac{r(z)}{\sigma(z)} = \sum_{j=0}^{l-1} \frac{\alpha^{i_j}}{(1 - \alpha^{i_j}z)}$$

$$= \sum_{j=0}^{l-1} \alpha^{i_j} \sum_{k=0}^{\infty} (\alpha^{i_j}z)^k$$

$$\equiv \sum_{j=0}^{l-1} \alpha^{i_j} \sum_{k=0}^{\delta-2} (\alpha^{i_j}z)^k$$

$$\equiv \sum_{k=0}^{\delta-2} \left(\sum_{j=0}^{l-1} (\alpha^{i_j})^{k+1} \right) z^k$$

$$\equiv \sum_{k=0}^{\delta-2} w(\alpha^{k+1})z^k$$

$$\equiv s(z) \pmod{z^{\delta-1}}.$$

As $r(1/\alpha^{i_j}) \ne 0$ for all j, we know that $\gcd(r(z), \sigma(z)) = 1$. This completes the proof.   □

From the above theorem, we find that, to determine the error locator polynomial $\sigma(z)$, it is sufficient to solve the polynomial congruence (8.8). This can be done by the *Euclidean algorithm* (see ref. [11]).

Step 3: Finding the roots of the error locator polynomial This is easy to do as we can search for all possible roots by evaluating $\sigma(z)$ at α^i, for all $i = 1, 2, \ldots$. After all the roots $\alpha^{i_1}, \ldots, \alpha^{i_l}$ of $\sigma(z)$ are found, we obtain the error polynomial (8.6).

We use an example to illustrate the above three steps.

Example 8.1.22 Let α be a root of $g(x) = 1 + x + x^3 \in \mathbf{F}_2[x]$. Then the Hamming code generated by $g(x) = \mathrm{lcm}(M^{(1)}(x), M^{(2)}(x))$ has the designed distance $\delta = 3$. Suppose that $w(x) = 1 + x + x^2 + x^3$ is a received word.

(i) Calculation of syndromes:

$$(s_0, s_1) = (w(\alpha), w(\alpha^2)) = (\alpha^2, \alpha^4).$$

(ii) Finding the error locator polynomial.
Solve the polynomial congruence

$$r(z) \equiv s(z)\sigma(z) \,(\mathrm{mod}\ z^2)$$

with $\deg(r(z)) = 0$ and $\deg(\sigma(z)) \leq 1$, and

$$s(z) = \alpha^2 + \alpha^4 z.$$

We have $\sigma(z) = 1 + \alpha^2 z$ and $r(z) = \alpha^2$. Hence, the error takes place at the third position. Thus, we decode $w(x)$ to $w(x) - x^2 = 1 + x + x^3 = 1101000$.

8.2   Reed–Solomon codes

The most important subclass of BCH codes is the class of Reed–Solomon (RS) codes. RS codes were introduced by I. S. Reed and G. Solomon independently of the work by A. Hocquenghem, R. C. Bose and D. K. Ray-Chaudhuri.

Consider a q-ary BCH code C of length $q^m - 1$ generated by $g(x) :=$ $\mathrm{lcm}(M^{(a)}(x), M^{(a+1)}(x), \ldots, M^{(a+\delta-2)}(x))$, where $M^{(i)}(x)$ is the minimal polynomial of α^i with respect to $\mathbf{F}_q$ for a primitive element α of $\mathbf{F}_{q^m}$. If $m = 1$, we obtain a q-ary BCH code of length $q - 1$. In this case, α is a primitive element of $\mathbf{F}_q$ and, moreover, the minimal polynomial of α^i with respect to $\mathbf{F}_q$ is $x - \alpha^i$. Thus, for $\delta \leq q - 1$, the generator polynomial becomes

$$\begin{aligned} g(x) &= \mathrm{lcm}(x - \alpha^a, x - \alpha^{a+1}, \ldots, x - a^{a+\delta-2}) \\ &= (x - \alpha^a)(x - \alpha^{a+1}) \cdots (x - a^{a+\delta-2}) \end{aligned}$$

since $\alpha^a, \alpha^{a+1}, \ldots, a^{a+\delta-2}$ are pairwise distinct.

Definition 8.2.1 A q-ary *Reed–Solomon code (RS code)* is a q-ary BCH code of length $q - 1$ generated by

$$g(x) = (x - \alpha^{a+1})(x - \alpha^{a+2}) \cdots (x - \alpha^{a+\delta-1}),$$

with $a \geq 0$ and $2 \leq \delta \leq q - 1$, where α is a primitive element of $\mathbf{F}_q$.

We never consider binary RS codes as, in this case, the length is $q - 1 = 1$.

Example 8.2.2 (i) Consider the 7-ary RS code of length 6 with generator polynomial $g(x) = (x - 3)(x - 3^2)(x - 3^3) = 6 + x + 3x^2 + x^3$. This is a 7-ary [6, 3]-code.

We can form a generator matrix from $g(x)$:

$$G = \begin{pmatrix} 6 & 1 & 3 & 1 & 0 & 0 \\ 0 & 6 & 1 & 3 & 1 & 0 \\ 0 & 0 & 6 & 1 & 3 & 1 \end{pmatrix}.$$

A parity-check matrix

$$H = \begin{pmatrix} 1 & 4 & 1 & 1 & 0 & 0 \\ 0 & 1 & 4 & 1 & 1 & 0 \\ 0 & 0 & 1 & 4 & 1 & 1 \end{pmatrix}$$

is obtained from $h(x) = (x^6 - 1)/g(x) = 1 + x + 4x^2 + x^3$. It can be checked from the above parity-check matrix that the minimum distance is 4. Hence, this is a 7-ary $[6, 3, 4]$-MDS code.

(ii) Consider the 8-ary RS code of length 7 with generator polynomial $g(x) = (x - \alpha)(x - \alpha^2) = 1 + \alpha + (\alpha^2 + \alpha)x + x^2$, where α is a root of $1 + x + x^3 \in F_2[x]$. This is an 8-ary $[7, 5]$-code.

We can form a generator matrix from $g(x)$:

$$G = \begin{pmatrix} \alpha + 1 & \alpha^2 + \alpha & 1 & 0 & 0 & 0 & 0 \\ 0 & \alpha + 1 & \alpha^2 + \alpha & 1 & 0 & 0 & 0 \\ 0 & 0 & \alpha + 1 & \alpha^2 + \alpha & 1 & 0 & 0 \\ 0 & 0 & 0 & \alpha + 1 & \alpha^2 + \alpha & 1 & 0 \\ 0 & 0 & 0 & 0 & \alpha + 1 & \alpha^2 + \alpha & 1 \end{pmatrix}.$$

A parity-check matrix

$$H = \begin{pmatrix} 1 & \alpha^4 & 1 & 1 + \alpha^4 & 1 + \alpha^4 & \alpha^4 & 0 \\ 0 & 1 & \alpha^4 & 1 & 1 + \alpha^4 & 1 + \alpha^4 & \alpha^4 \end{pmatrix}$$

is obtained from $h(x) = (x^7 - 1)/g(x) = \alpha^4 + (1 + \alpha^4)x + (1 + \alpha^4)x^2 + x^3 + \alpha^4 x^4 + x^5$. It can be checked from the above parity-check matrix that the minimum distance is 3. Hence, this is an 8-ary $[7, 5, 3]$-MDS code.

The two RS codes in the above example are both MDS. In fact, it is true in general that RS codes are MDS.

Theorem 8.2.3 *Reed–Solomon codes are MDS; i.e., a q-ary Reed–Solomon code of length $q - 1$ generated by $g(x) = \prod_{i=a+1}^{a+\delta-1}(x - \alpha^i)$ is a $[q - 1, q - \delta, \delta]$-cyclic code for any $2 \le \delta \le q - 1$.*

Proof. As the degree of $g(x)$ is $\delta - 1$, the dimension of the code is exactly $k := q - 1 - (\delta - 1) = q - \delta$.

By Theorem 8.1.18, the minimum distance is at least δ. On the other hand, the minimum distance is at most $(q - 1) + 1 - k = \delta$ by the Singleton bound (see Theorem 5.4.1). The desired result follows.   □

Example 8.2.4 Let α be a root of $1 + x + x^4 \in \mathbf{F}_2[x]$. Then α is a primitive element of $\mathbf{F}_{16}$. Consider the RS code generated by $g(x) = (x - \alpha^3)$ $(x - \alpha^4)(x - \alpha^5)(x - \alpha^6)$. It is not so easy to work out its minimum distance from its parity-check matrices. However, by Theorem 8.2.3, it is a $[15, 11, 5]$-cyclic code over $\mathbf{F}_{16}$.

Next we consider the extended codes of RS codes.

Theorem 8.2.5 *Let C be a q-ary RS code generated by $g(x) = \prod_{i=1}^{\delta-1}(x - \alpha^i)$ with $2 \le \delta \le q - 1$. Then the extended code $\overline{C}$ is still MDS.*

Proof. Since C is a $[q - 1, q - \delta, \delta]$-cyclic code, we have to show that $\overline{C}$ is a $[q, q - \delta, \delta + 1]$-code. Let $c(x) = \sum_{i=0}^{q-2} c_i x^i$ be a nonzero codeword of C. It is sufficient to prove that the Hamming weight of $\overline{c} = (c_0, \ldots, c_{q-2}, - \sum_{i=0}^{q-2} c_i)$ is at least $\delta + 1$. Let $c(x) = f(x)g(x)$ for some $f(x) \in \mathbf{F}_q[x]/ (x^{q-1} - 1)$.

Case 1: $f(1) \ne 0$. It is clear that $g(1) \ne 0$. Hence, $c(1) = \sum_{i=0}^{q-2} c_i \ne 0$. Then the Hamming weight of $\overline{c}$ is equal to $\mathrm{wt}(c(x)) + 1$, which is at least $d(C) + 1 = \delta + 1$.

Case 2: $f(1) = 0$, i.e., $(x - 1)$ is a linear factor of $f(x)$. Put $f(x) = (x - 1)u(x)$. Then, $c(x) = u(x)(x - 1)g(x) = u(x)\prod_{i=0}^{\delta-1}(x - \alpha^i)$ is also a codeword of the BCH code of designed distance $\delta + 1$ generated by $\prod_{i=0}^{\delta-1}(x - \alpha^i)$. Hence, the Hamming weight of $c(x)$ is at least $\delta + 1$ by Theorem 8.1.18. Thus, the Hamming weight of $\overline{c}$ is at least $\delta + 1$.   □

Example 8.2.6 (i) Consider the 7-ary $[6, 3, 4]$-RS code as in Example 8.2.2(i). By Theorem 5.1.9, the matrix

$$\begin{pmatrix} 1 & 4 & 1 & 1 & 0 & 0 & 0 \\ 0 & 1 & 4 & 1 & 1 & 0 & 0 \\ 0 & 0 & 1 & 4 & 1 & 1 & 0 \\ 1 & 1 & 1 & 1 & 1 & 1 & 1 \end{pmatrix}$$

is a parity-check matrix of the extended code. Hence, by Corollary 4.5.7, the extended code has minimum distance 5, and thus it is a $[7, 3, 5]$-MDS code.

(ii) Consider the 8-ary [7, 5, 3]-RS code as in Example 8.2.2(ii). By Theorem 5.1.9, the matrix

$$\begin{pmatrix} 1 & \alpha^4 & 1 & 1+\alpha^4 & 1+\alpha^4 & \alpha^4 & 0 & 0 \\ 0 & 1 & \alpha^4 & 1 & 1+\alpha^4 & 1+\alpha^4 & \alpha^4 & 0 \\ 1 & 1 & 1 & 1 & 1 & 1 & 1 & 1 \end{pmatrix}$$

is a parity-check matrix of the extended code. Hence, the extended code has minimum distance 4, and thus it is an [8, 5, 4]-MDS code.

RS codes are MDS codes. Hence, they have very good parameters. Unfortunately, RS codes are nonbinary, while practical applications often require binary codes. In practice, the concatenation technique is used to produce binary codes from RS codes over extension fields of $\mathbf{F}_2$.

Let C be an $[n, k]$-RS code over $\mathbf{F}_{2^m}$, where $n = 2^m - 1$. Applying the concatenation technique as in Theorem 6.3.1, we concatenate C with the trivial code $\mathbf{F}_2^m$.

Let $\alpha_1, \ldots, \alpha_m$ be an $\mathbf{F}_2$-basis of $\mathbf{F}_{2^m}$ and consider the map $\phi : \mathbf{F}_{2^m} \to \mathbf{F}_2^m$

$$u_1\alpha_1 + u_2\alpha_2 + \cdots + u_m\alpha_m \mapsto (u_1, u_2, \ldots, u_m).$$

Then, by Theorem 6.3.1, we have the following result.

Theorem 8.2.7 *Let C be an $[n, k]$-RS code over $\mathbf{F}_{2^m}$, where $n = 2^m - 1$. Then*

$$\phi^*(C) := \{(\phi(c_0), \ldots, \phi(c_{n-1})) : (c_0, \ldots, c_{n-1}) \in C\}$$

is a binary $[mn, mk]$-code with minimum distance at least $n - k + 1$.

Example 8.2.8 Consider the 8-ary RS code C generated by

$$g(x) = \prod_{i=1}^{6}(x - \alpha^i) = \sum_{i=0}^{6} x^i,$$

where α is a root of $1 + x + x^3$. Hence,

$$C = \{a(1, 1, 1, 1, 1, 1, 1) : a \in \mathbf{F}_8\}$$

is the trivial 8-ary [7, 1, 7]-MDS code. The code $\phi^*(C)$ is a binary [21, 3, 7]-linear code spanned by

$$100\,100 \ldots 100, \quad 010\,010 \ldots 010, \quad 001\,001 \ldots 001.$$

For an RS code C, the code $\phi^*(C)$ cannot correct too many random errors as the minimum distance is not very big. However, it can correct many more burst errors.

Theorem 8.2.9 *Let C be an $[2^m - 1, k]$-RS code over $\mathbf{F}_{2^m}$. Then, the code $\phi^*(C)$ can correct $m\lfloor (n - k)/2 \rfloor - m + 1$ burst errors, where $n = 2^m - 1$ is the length of the code.*

Proof. Put $l = m\lfloor (n - k)/2 \rfloor - m + 1$. By Theorem 7.5.3, it is sufficient to show that all the burst errors of length l or less lie in distinct cosets.

Let $\mathbf{e}_1, \mathbf{e}_2 \in \mathbf{F}_2^{mn}$ be two burst errors of length l or less that lie in the same coset of $\phi^*(C)$. Let $\mathbf{c}_i$ be the pre-image of $\mathbf{e}_i$ under the map ϕ^*; i.e., $\phi^*(\mathbf{c}_i) = \mathbf{e}_i$ for $i = 1, 2$. Then it is clear that

$$\begin{aligned}
\text{wt}(\mathbf{c}_i) &\le \left\lceil \frac{l - 1}{m} \right\rceil + 1 \\
&= \left\lfloor \frac{n - k}{2} \right\rfloor \\
&= \left\lfloor \frac{d(C) - 1}{2} \right\rfloor \quad \text{(as C is an MDS code),}
\end{aligned}$$

for $i = 1, 2$, and $\mathbf{c}_1, \mathbf{c}_2$ are in the same coset of C. By Exercise 4.44, we know that $\mathbf{c}_1 = \mathbf{c}_2$. This means that $\mathbf{e}_1 = \mathbf{e}_2$ since ϕ^* is injective.   □

Example 8.2.10 For an 8-ary $[7, 3, 5]$-RS code, the code $\phi^*(C)$ is a binary $[21, 9]$-linear code. It can correct

$$l = 3 \left\lfloor \frac{7 - 3}{2} \right\rfloor - 3 + 1 = 4$$

burst errors.

8.3   Quadratic-residue codes

Quadratic-residue (QR) codes have been extensively studied for many years. Examples of good quadratic-residue codes are the binary $[7, 4, 3]$-Hamming code, the binary $[23, 12, 7]$-Golay code and the ternary $[11, 6, 5]$-Golay code.

Let p be a prime number bigger than 2 and choose a primitive element g of $\mathbf{F}_p$ (we know the existence of primitive elements by Proposition 3.3.9(ii)). A nonzero element r of $\mathbf{F}_p$ is called a *quadratic residue modulo p* if $r = g^{2i}$ for some integer i when r is viewed as an element of $\mathbf{F}_p$; otherwise, r is called a *quadratic nonresidue modulo p*. It is clear that r is a quadratic nonresidue modulo p if and only if $r = g^{2j-1}$ for some integer j.

Example 8.3.1 (i) Consider the finite field $\mathbf{F}_7$. It is easy to check that 3 is a primitive element of $\mathbf{F}_7$. Thus, the nonzero quadratic residues modulo 7 are

$\{3^{2i} : i = 0, 1, \ldots\} = \{1, 2, 4\}$, and the quadratic nonresidues modulo 7 are $\{3^{2i-1} : i = 1, 2, \ldots\} = \{3, 6, 5\}$.

(ii) Consider the finite field $\mathbf{F}_{11}$. It is easy to check that 2 is a primitive element of $\mathbf{F}_{11}$. Thus, the nonzero quadratic residues modulo 11 are $\{2^{2i} : i = 0, 1, \ldots\} = \{1, 4, 5, 9, 3\}$, and the quadratic nonresidues modulo 11 are $\{2^{2i-1} : i = 1, 2, \ldots\} = \{2, 8, 10, 7, 6\}$.

(iii) Consider the finite field $\mathbf{F}_{23}$. It is easy to check that 5 is a primitive element of $\mathbf{F}_{23}$. Thus, the nonzero quadratic residues modulo 23 are $\{5^{2i} : i = 0, 1, \ldots\} = \{1, 2, 4, 8, 16, 9, 18, 13, 3, 6, 12\}$, and the quadratic nonresidues modulo 23 are $\{5^{2i-1} : i = 1, 2, \ldots\} = \{5, 10, 20, 17, 11, 22, 21, 19, 15, 7, 14\}$.

We now show that quadratic residues modulo p are independent of the choice of the primitive element.

Proposition 8.3.2 *A nonzero element r of $\mathbf{F}_p$ is a nonzero quadratic residue modulo p if and only if $r \equiv a^2 \pmod{p}$ for some $a \in \mathbf{F}_p^*$. In particular, quadratic residues modulo p are independent of the choice of the primitive element.*

Proof. Let g be a primitive element of $\mathbf{F}_p$. If r is a nonzero quadratic residue modulo p, then, by the definition, $r = g^{2i}$ for some integer i. Putting $a = g^i$, we have $r \equiv a^2 \pmod{p}$.

Conversely, if $r \equiv a^2 \pmod{p}$ for some $a \in \mathbf{F}_p^*$, then $r = a^2$ in $\mathbf{F}_p$. Since g is a primitive element of $\mathbf{F}_p$, there exists an integer i such that $a = g^i$. Thus, $r = g^{2i}$; i.e., r is a quadratic residue modulo p.   □

Example 8.3.3 2 is a quadratic residue modulo 17 as $2 \equiv 6^2 \pmod{17}$.

Proposition 8.3.4 *Let p be an odd prime. Denote by $\mathcal{Q}_p$ and $\mathcal{N}_p$ the sets of nonzero quadratic residues and quadratic nonresidues modulo p, respectively. Then we have the following.*

(i) *The product of two quadratic residues modulo p is a quadratic residue modulo p.*

(ii) *The product of two quadratic nonresidues modulo p is a quadratic residue modulo p.*

(iii) *The product of a nonzero quadratic residue modulo p with a quadratic nonresidue modulo p is a quadratic nonresidue modulo p.*

(iv) *There are exactly $(p-1)/2$ nonzero quadratic residues modulo p and $(p-1)/2$ quadratic nonresidues modulo p, and therefore $\mathbf{F}_p = \{0\} \cup \mathcal{Q}_p \cup \mathcal{N}_p$.*

(v) *For $\alpha \in \mathcal{Q}_p$ and $\beta \in \mathcal{N}_p$, we have that*

$$\alpha \mathcal{Q}_p = \{\alpha r : r \in \mathcal{Q}_p\} = \mathcal{Q}_p,$$
$$\beta \mathcal{Q}_p = \{\beta r : r \in \mathcal{Q}_p\} = \mathcal{N}_p,$$
$$\alpha \mathcal{N}_p = \{\alpha n : n \in \mathcal{N}_p\} = \mathcal{N}_p$$

and

$$\beta \mathcal{N}_p = \{\beta n : n \in \mathcal{N}_p\} = \mathcal{Q}_p.$$

Proof. Let g be a primitive element of $\mathbf{F}_p$, and let γ, θ be two quadratic residues modulo p. Then, there exist two integers i, j such that $\gamma = g^{2i}$ and $\theta = g^{2j}$. Hence, $\gamma \theta = g^{2(i+j)}$ is a quadratic residue modulo p.

The same arguments can be employed to prove parts (ii) and (iii).

It is clear that all the nonzero quadratic residues modulo p are

$$\{g^{2i} : i = 0, 1, \ldots, (p-3)/2\},$$

and that all the quadratic nonresidues modulo p are

$$\{g^{2i-1} : i = 1, 2, \ldots, (p-1)/2\}.$$

Thus, part (iv) follows.

Part (v) follows from parts (i)–(iv) immediately. $\square$

Example 8.3.5 Consider the finite field $\mathbf{F}_{11}$. The set of nonzero quadratic residues modulo 11 is $\mathcal{Q}_{11} = \{1, 4, 5, 9, 3\}$, and the set of quadratic nonresidues modulo 11 is $\mathcal{N}_{11} = \{2, 8, 10, 7, 6\}$. We have $|\mathcal{Q}_{11}| = |\mathcal{N}_{11}| = 5 = (11-1)/2$. Furthermore, by choosing $4 \in \mathcal{Q}_{11}$ and $2 \in \mathcal{N}_{11}$, we have

$$4\mathcal{Q}_{11} = \{4 \cdot 1, \ 4 \cdot 4, \ 4 \cdot 5, \ 4 \cdot 9, \ 4 \cdot 3\} = \{4, 5, 9, 3, 1\} = \mathcal{Q}_{11},$$
$$2\mathcal{Q}_{11} = \{2 \cdot 1, \ 2 \cdot 4, \ 2 \cdot 5, \ 2 \cdot 9, \ 2 \cdot 3\} = \{2, 8, 10, 7, 6\} = \mathcal{N}_{11},$$
$$4\mathcal{N}_{11} = \{4 \cdot 2, \ 4 \cdot 8, \ 4 \cdot 10, \ 4 \cdot 7, \ 4 \cdot 6\} = \{8, 10, 7, 6, 2\} = \mathcal{N}_{11}$$

and

$$2\mathcal{N}_{11} = \{2 \cdot 2, \ 2 \cdot 8, \ 2 \cdot 10, \ 2 \cdot 7, \ 2 \cdot 6\} = \{4, 5, 9, 3, 1\} = \mathcal{Q}_{11}.$$

Choose a prime l such that $l \neq p$ and l is a quadratic residue modulo p. Choose an integer $m \geq 1$ such that $l^m - 1$ is divisible by p. Let θ be a primitive element of $\mathbf{F}_{l^m}$ and put $\alpha = \theta^{(l^m-1)/p}$. Then, the order of α is p; i.e., $1 = \alpha^0 = \alpha^p$, $\alpha = \alpha^1, \alpha^2, \ldots, \alpha^{p-1}$ are pairwise distinct and $x^p - 1 = \prod_{i=0}^{p-1}(x - \alpha^i)$.

Consider the polynomials

$$g_Q(x) := \prod_{r \in \mathcal{Q}_p}(x - \alpha^r) \quad \text{and} \quad g_N(x) := \prod_{n \in \mathcal{N}_p}(x - \alpha^n). \qquad (8.12)$$

It follows from Proposition 8.3.4(iv) that

$$x^p - 1 = (x - 1)g_Q(x)g_N(x). \tag{8.13}$$

Moreover, we have the following result.

Lemma 8.3.6 *The polynomials $g_Q(x)$ and $g_N(x)$ belong to $\mathbf{F}_l[x]$.*

Proof. It is sufficient to show that each coefficient of $g_Q(x)$ and $g_N(x)$ belongs to $\mathbf{F}_l$.

Let $g_Q(x) = a_0 + a_1 x + \cdots + a_k x^k$, where $a_i \in \mathbf{F}_{l^m}$ and $k = (p-1)/2$. Raising each coefficient to its lth power, we obtain

$$
\begin{aligned}
a_0^l + a_1^l x + \cdots + a_k^l x^k &= \prod_{r \in \mathcal{Q}_p} (x - \alpha^{lr}) \\
&= \prod_{j \in l\mathcal{Q}_p} (x - \alpha^j) \\
&= \prod_{j \in \mathcal{Q}_p} (x - \alpha^j) \\
&= g_Q(x).
\end{aligned}
$$

Note that we use the fact that $l\mathcal{Q}_p = \mathcal{Q}_p$ in the above argument. Hence, $a_i = a_i^l$ for all $0 \le i \le m$; i.e., a_i are elements of $\mathbf{F}_l$. This means that $g_Q(x)$ is a polynomial over $\mathbf{F}_l$.

The same argument can be used to show that $g_N(x)$ is a polynomial over $\mathbf{F}_l$.   □

Example 8.3.7 (i) Let $p = 7$ and $l = 2$. Let α be a root of $1 + x + x^3 \in \mathbf{F}_2[x]$. Then the order of α is 7. The two polynomials defined in (8.12) are

$$
\begin{aligned}
g_Q(x) &= \prod_{r \in \mathcal{Q}_7} (x - \alpha^r) \\
&= (x - \alpha)(x - \alpha^2)(x - \alpha^4) \\
&= 1 + x + x^3
\end{aligned}
$$

and

$$
\begin{aligned}
g_N(x) &= \prod_{n \in \mathcal{N}_7} (x - \alpha^n) \\
&= (x - \alpha^3)(x - \alpha^6)(x - \alpha^5) \\
&= 1 + x^2 + x^3.
\end{aligned}
$$

Furthermore, we have

$$x^7 - 1 = (x - 1)g_Q(x)g_N(x).$$

(ii) Let $p = 11$ and $l = 3$. Let θ be a root of $1 + 2x + x^5 \in \mathbf{F}_3[x]$. Then θ is a primitive element of $\mathbf{F}_{3^5}$, and the order of $\alpha := \theta^{22}$ is 11. The two polynomials defined in (8.12) are

$$g_Q(x) = \prod_{r \in \mathcal{Q}_{11}} (x - \alpha^r)$$

$$= (x - \alpha)(x - \alpha^4)(x - \alpha^5)(x - \alpha^9)(x - \alpha^3)$$

$$= 2 + x^2 + 2x^3 + x^4 + x^5$$

and

$$g_N(x) = \prod_{n \in \mathcal{N}_{11}} (x - \alpha^n)$$

$$= (x - \alpha^2)(x - \alpha^8)(x - \alpha^{10})(x - \alpha^7)(x - \alpha^6)$$

$$= 2 + 2x + x^2 + 2x^3 + x^5.$$

Furthermore, we have

$$x^{11} - 1 = (x - 1)g_Q(x)g_N(x).$$

(iii) Let $p = 23$ and $l = 2$. Let θ be a root of $1 + x + x^3 + x^5 + x^{11} \in \mathbf{F}_2[x]$. Then θ is a primitive element of $\mathbf{F}_{2^{11}}$, and the order of $\alpha := \theta^{89}$ is 23. The two polynomials defined in (8.12) are

$$g_Q(x) = \prod_{r \in \mathcal{Q}_{23}} (x - \alpha^r)$$

$$= (x - \alpha)(x - \alpha^2)(x - \alpha^4)(x - \alpha^8)(x - \alpha^{16})$$

$$\times (x - \alpha^9)(x - \alpha^{18})(x - \alpha^{13})(x - \alpha^3)(x - \alpha^6)(x - \alpha^{12})$$

$$= 1 + x^2 + x^4 + x^5 + x^6 + x^{10} + x^{11}$$

and

$$g_N(x) = \prod_{n \in \mathcal{N}_{23}} (x - \alpha^n)$$

$$= (x - \alpha^5)(x - \alpha^{10})(x - \alpha^{20})(x - \alpha^{17})(x - \alpha^{11})$$

$$\times (x - \alpha^{22})(x - \alpha^{21})(x - \alpha^{19})(x - \alpha^{15})(x - \alpha^7)(x - \alpha^{14})$$

$$= 1 + x + x^5 + x^6 + x^7 + x^9 + x^{11}.$$

Furthermore, we have

$$x^{23} - 1 = (x - 1)g_Q(x)g_N(x).$$

Definition 8.3.8 Let p and l be two distinct primes such that l is a quadratic residue modulo p. Choose an integer $m \geq 1$ such that $l^m - 1$ is divisible by p. Let θ be a primitive element of $\mathbf{F}_{l^m}$ and put $\alpha = \theta^{(l^m-1)/p}$. The divisors of $x^p - 1$

$$g_Q(x) := \prod_{r \in \mathcal{Q}_p}(x - \alpha^r) \quad \text{and} \quad g_N(x) := \prod_{n \in \mathcal{N}_p}(x - \alpha^n)$$

are defined over $\mathbf{F}_l$. The l-ary cyclic codes $C_Q =< g_Q(x) >$ and $C_N = < g_N(x) >$ of length p are called *quadratic-residue (QR) codes.*

It is obvious that the dimensions of both the quadratic-residue codes C_Q and C_N are $p - (p-1)/2 = (p+1)/2$.

Example 8.3.9 (i) Consider the binary quadratic-residue codes $C_Q =< 1 + x + x^3 >$ and $C_N =< 1 + x^2 + x^3 >$ of length 7. It is easy to verify that these two codes are equivalent (see Proposition 8.3.12) and that both are binary $[7, 4, 3]$-Hamming codes.

(ii) Consider the ternary quadratic-residue codes $C_Q =< 2 + x^2 + 2x^3 + x^4 + x^5 >$ and $C_N =< 2 + 2x + x^2 + 2x^3 + x^5 >$ of length 11. It is easy to verify that these two codes are equivalent (see Proposition 8.3.12) and that both are equivalent to the ternary $[11, 6, 5]$-Golay code defined in Section 5.3.

(iii) Consider the binary quadratic-residue codes $C_Q =< 1 + x^2 + x^4 + x^5 + x^6 + x^{10} + x^{11} >$ and $C_N =< 1 + x + x^5 + x^6 + x^7 + x^9 + x^{11} >$ of length 23. It is easy to verify that these two codes are equivalent (see Proposition 8.3.12) and that both are equivalent to the binary $[23, 12, 7]$-Golay code defined in Section 5.3.

From the above example, we can see that the codes C_Q and C_N are equivalent in these three cases. This is, in fact, true in general. We prove the following lemma first.

Lemma 8.3.10 *Let m, n be two integers bigger than 1 and $\gcd(m, n) = 1$. Then the map*

$$\chi_m : \mathbf{F}_q[x]/(x^n - 1) \rightarrow \mathbf{F}_q[x]/(x^n - 1), \quad a(x) \mapsto a(x^m)$$

is a permutation of $\mathbf{F}_q^n$ if we identify $\mathbf{F}_q^n$ with $\mathbf{F}_q[x]/(x^n - 1)$ through the map

$$\pi : (f_0, f_1, \ldots, f_{n-1}) \mapsto \sum_{i=0}^{n-1} f_i x^i.$$

Proof. Let $f(x) = \sum_{i=0}^{n-1} f_i x^i$. Then, we have

$$\chi_m(f(x)) = f(x^m) \ (\text{mod } x^n - 1) = \sum_{i=0}^{n-1} f_i x^{(mi \ (\text{mod } n))}.$$

Hence, it is sufficient to show that

$$0, \quad (m \ (\text{mod } n)), \quad (2m \ (\text{mod } n)), \quad \ldots, \quad ((n-1)m \ (\text{mod } n))$$

is a permutation of $0, 1, 2, \ldots, n-1$. This is clearly true, as $\gcd(m, n) = 1$.

$\square$

Example 8.3.11 Consider the map

$$\chi_3 : \mathbf{F}_3[x]/(x^5 - 1) \rightarrow \mathbf{F}_3[x]/(x^5 - 1), \quad a(x) \mapsto a(x^3).$$

Then,

$$\chi_3(1 + 2x + x^4) = 1 + 2x^3 + x^{12} \ (\text{mod } x^5 - 1) = 1 + x^2 + 2x^3.$$

Clearly, $(1, 0, 1, 2, 0)$ is a permutation of $(1, 2, 0, 0, 1)$.

Proposition 8.3.12 *The two l-ary quadratic-residue codes C_Q and C_N are equivalent.*

Proof. By definition, $C_Q =< g_Q(x) >$ and $C_N =< g_N(x) >$. Choose a quadratic nonresidue m modulo p and consider the map

$$\chi_m : \mathbf{F}_l[x]/(x^p - 1) \rightarrow \mathbf{F}_l[x]/(x^p - 1), \quad a(x) \mapsto a(x^m).$$

Then, $\chi_m(C_Q)$ is an equivalent code of C_Q by Lemma 8.3.10. We claim that the code $\chi_m(C_Q)$ is in fact the same as C_N. This is equivalent to $\chi_m(C_Q) \subseteq C_N$ as $|\chi_m(C_Q)| = |C_N|$. Hence, it is sufficient to show that $\chi_m(g_Q(x)) \in C_N$; i.e., $g_N(x) = \prod_{t \in N_p}(x - \alpha^t)$ is a divisor of $\chi_m(g_Q(x)) = \prod_{r \in Q_p}(x^m - \alpha^r)$. Let t be a quadratic nonresidue modulo p, then tm is a nonzero quadratic residue modulo p. Hence,

$$0 = g_Q(\alpha^{tm}) = g_Q((\alpha^t)^m) = \chi_m(g_Q(\alpha^t)).$$

This implies that $g_N(x)$ is a divisor of $\chi_m(g_Q(x))$ as $g_N(x)$ has no multiple roots.

$\square$

Finally, we determine the possible lengths p for which a binary quadratic-residue code exists; i.e., those primes p such that 2 is a quadratic residue modulo p.

Table 8.3.

Length	Dimension	Distance
7	4	3
17	9	5
23	12	7
31	16	7
41	21	9
47	24	11
71	36	11
73	37	13
79	40	15
89	45	17

Proposition 8.3.13 (i) *Let p be an odd prime and let r be an integer such that $\gcd(r, p) = 1$. Then, r is a quadratic residue modulo p if and only if $r^{(p-1)/2} \equiv 1 \pmod{p}$.*

(ii) *For an odd prime p, 2 is a quadratic residue modulo p if p is of the form $p = 8m \pm 1$, and it is a quadratic nonresidue modulo p if p is of the form $p = 8m \pm 3$.*

Proof. (i) Let g be a primitive element of $\mathbf{F}_p$. If r is a quadratic residue modulo p, then $r = g^{2i}$ for some i. Hence, $r^{(p-1)/2} = g^{i(p-1)} = 1$ in $\mathbf{F}_p$; i.e., $r^{(p-1)/2} \equiv 1 \pmod{p}$.

Conversely, suppose that $r^{(p-1)/2} \equiv 1 \pmod{p}$. Let $r = g^j$ for some integer j. Then $g^{j(p-1)/2} = 1$ in $\mathbf{F}_p$. This means that $j(p-1)/2$ is divisible by $p-1$; i.e., j is even.

(ii) Consider the following $(p-1)/2$ numbers:

$$2 \times 1 \qquad\qquad 2 \times 2 \qquad \cdots \qquad 2 \times \lfloor(p-1)/4\rfloor$$
$$p - 2(\lfloor(p-1)/4\rfloor + 1) \quad p - 2(\lfloor(p-1)/4\rfloor + 2) \quad \cdots \quad p - 2((p-1)/2).$$

All of these $(p-1)/2$ numbers are between 1 and $(p-1)/2$ (both inclusive) and it is easy to verify that they are pairwise distinct. Thus, their product is equal to

$$\left(\frac{p-1}{2}\right)! = \prod_{i=1}^{\lfloor(p-1)/4\rfloor} 2i \prod_{j=\lfloor(p-1)/4\rfloor+1}^{(p-1)/2} (p - 2j)$$

$$\equiv (-1)^e 2^{(p-1)/2} \left(\frac{p-1}{2}\right)! \pmod{p},$$

where $e = (p-1)/2 - \lfloor (p-1)/4 \rfloor$. Hence, we obtain $2^{(p-1)/2} \equiv (-1)^e \pmod{p}$. It is easy to check that e is even if and only if p is of the form $p = 8m \pm 1$, and that e is odd if and only if p is of the form $p = 8m \pm 3$. The desired result then follows from part (i). $\qquad\square$

Corollary 8.3.14 *There exist binary quadratic-residue codes of length p if and only if p is a prime of the form $p = 8m \pm 1$.*

Example 8.3.15 We list the parameters of the first ten binary quadratic-residue codes in Table 8.3.

Exercises

8.1 Find the least common multiple of the following polynomials over $\mathbf{F}_2$:

$$f_1(x) = 1 + x^2, \quad f_2(x) = 1 + x + x^2 + x^4, \quad f_3(x) = 1 + x^2 + x^4 + x^6.$$

8.2 Suppose we have three nonzero polynomials $f_1(x)$, $f_2(x)$ and $f_3(x)$. Show that $\mathrm{lcm}(f_1(x), f_2(x), f_3(x)) = \mathrm{lcm}(\mathrm{lcm}(f_1(x), f_2(x)), f_3(x))$.

8.3 Construct a generator polynomial and a parity-check matrix for a binary double-error-correcting BCH code of length 15.

8.4 Let α be a root of $1 + x + x^4 \in \mathbf{F}_2[x]$.

    (a) Show that α^7 is a primitive element of $\mathbf{F}_{16}$, and find the minimal polynomial of α^7 with respect to $\mathbf{F}_2$.

    (b) Let $g(x) \in \mathbf{F}_2[x]$ be the polynomial of lowest degree such that $g(\alpha^{7i}) = 0$, for $i = 1, 2, 3, 4$. Determine $g(x)$ and construct a parity-check matrix of the binary cyclic code generated by $g(x)$.

8.5 Determine the generator polynomials of all binary BCH codes of length 31 with designed distance 5.

8.6 Construct the generator polynomial for a self-orthogonal binary BCH code of length 31 and dimension 15.

8.7 Let α be a root of $1 + x + x^4 \in \mathbf{F}_2[x]$. Let C be the narrow-sense binary BCH code of length 15 with designed distance 5.

    (a) Find the generator polynomial of C.

    (b) If possible, determine the error positions of the following received words:

        (i) $w(x) = 1 + x^6 + x^7 + x^8$;

        (ii) $w(x) = 1 + x + x^4 + x^5 + x^6 + x^9$;

        (iii) $w(x) = 1 + x + x^7$.

8.8 Let α be a root of $1 + x + x^4 \in F_2[x]$. Let C be the narrow-sense binary BCH code of length 15 with designed distance 7.

(a) Show that C is generated by $g(x) = 1 + x + x^2 + x^4 + x^5 + x^8 + x^{10}$.

(b) Let $w(x) = 1 + x + x^6 + x^7 + x^8$ be a received word. Find the syndrome polynomial, the error locator polynomial and decode the word $w(x)$.

8.9 Let C be a narrow-sense q-ary BCH code of length $n = q^m - 1$ with designed distance δ generated by $g(x) := \mathrm{lcm}(M^{(1)}(x),$ $M^{(2)}(x), \ldots, M^{(\delta-1)}(x))$, where $M^{(i)}(x)$ is the minimal polynomial of α^i with respect to F_q for a primitive element α of F_{q^m}.

Put

$$
H = \begin{pmatrix}
1 & \alpha & (\alpha)^2 & \cdots & (\alpha)^{n-1} \\
1 & \alpha^2 & (\alpha^2)^2 & \cdots & (\alpha^2)^{n-1} \\
1 & \alpha^3 & (\alpha^3)^2 & \cdots & (\alpha^3)^{n-1} \\
\cdot & \cdot & \cdot & & \cdot \\
\cdot & \cdot & \cdot & \cdots & \cdot \\
\cdot & \cdot & \cdot & & \cdot \\
1 & \alpha^{\delta-1} & (\alpha^{\delta-1})^2 & \cdots & (\alpha^{\delta-1})^{n-1}
\end{pmatrix}.
$$

Define the syndrome $S_H(\mathbf{w})$ of a word $\mathbf{w} \in F_q^n$ with respect to H by $\mathbf{w}H^T$. Show that, for any two words $\mathbf{u}, \mathbf{v} \in F_q^n$, we have

(a) $S_H(\mathbf{u} + \mathbf{v}) = S_H(\mathbf{u}) + S_H(\mathbf{v})$;

(b) $S_H(\mathbf{u}) = \mathbf{0}$ if and only if $\mathbf{u} \in C$;

(c) $S_H(\mathbf{u}) = S_H(\mathbf{v})$ if and only if $\mathbf{u}$ and $\mathbf{v}$ are in the same coset of C.

8.10 Show that the minimum distance of a narrow-sense binary BCH code is always odd.

8.11 Show that a narrow-sense binary BCH code of length $n = 2^m - 1$ and designed distance $2t + 1$ has minimum distance $2t + 1$, provided that

$$
\sum_{i=0}^{t+1} \binom{2^m - 1}{i} > 2^{mt}.
$$

8.12 Show that the narrow-sense binary BCH codes of length 31 and designed distance $\delta = 3, 5, 7$ have minimum distance 3, 5, 7, respectively.

8.13 Show that the minimum distance of a q-ary BCH code of length n and designed distance δ is equal to δ, provided that n is divisible by δ.

8.14 (i) Show that the cyclotomic cosets $C_1, C_3, C_5, \ldots, C_{2t+1}$ of 2 modulo $2^m - 1$ are pairwise distinct and that each contains exactly m elements, provided

$$
2t + 1 < 2^{\lfloor m/2 \rfloor} + 1.
$$

(ii) Show that a narrow-sense binary BCH code of length $n = 2^m - 1$ with designed distance $2t + 1$ has dimension $n - mt$ if

$$2t + 1 < 2^{\lfloor m/2 \rfloor} + 1.$$

8.15 Determine whether the dual of an arbitrary BCH code is a BCH code.

8.16 Find a generator matrix of a $[10, 6]$-RS code over $\mathbf{F}_{11}$ and determine the minimum distance.

8.17 Determine the generator polynomial of a 16-ary RS code of dimension 10 and find a parity-check matrix.

8.18 Show that, for all $n \le q$ and $1 \le k \le n$, there exists an $[n, k]$-MDS code over $\mathbf{F}_q$.

8.19 Show that the dual of an RS code is again an RS code.

8.20 Determine the generator polynomials of all the 16-ary self-orthogonal RS codes.

8.21 Let α be a root of $1 + x + x^2 \in \mathbf{F}_2[x]$. Consider the map

$$\phi : \mathbf{F}_{2^2} \to \mathbf{F}_2^2, \qquad a_0 + a_1\alpha \mapsto (a_0, a_1).$$

Let C be a $[3, 2]$-RS code over $\mathbf{F}_4$. Determine the parameters of $\phi^*(C)$.

8.22 Let C be a q-ary RS code generated by $g(x) = \prod_{i=1}^{\delta-1}(x - \alpha^i)$ with $3 \le \delta \le q - 2$, where α is a primitive element of $\mathbf{F}_q$. Show that the extended code $\overline{C}$ is equivalent to a cyclic code if and only if q is a prime.

8.23 Determine all quadratic residues modulo $p = 17, 29, 31$, respectively.

8.24 For an odd prime p, define Legendre's symbol by

$$\left(\frac{a}{p}\right) = \begin{cases} 0 & \text{if } p|a \\ 1 & \text{if } a \text{ is a quadratic residue and } \gcd(a, p) = 1 \\ -1 & \text{if } a \text{ is a quadratic nonresidue.} \end{cases}$$

(a) Show that

$$a^{(p-1)/2} \equiv \left(\frac{a}{p}\right) \pmod{p}.$$

(b) Show that

$$\left(\frac{a}{p}\right)\left(\frac{b}{p}\right) = \left(\frac{ab}{p}\right).$$

(c) Show that, if q is an odd prime and $p \ne q$, then

$$\left(\frac{q}{p}\right)\left(\frac{p}{q}\right) = (-1)^{(p-1)(q-1)/4}.$$

(Note: this is the law of *quadratic reciprocity*.)

8.25 For the following primes p, l and elements α, determine the polynomials $g_Q(x)$ and $g_N(x)$ over $\mathbf{F}_l$ as defined in Definition 8.3.8.
   (a) $p = 7, l = 2$ and α is a root of $1 + x^2 + x^3 \in \mathbf{F}_2[x]$.
   (b) $p = 17, l = 2$ and α is a root of $1 + x^2 + x^3 + x^4 + x^8 \in \mathbf{F}_2[x]$.
   (c) $p = 13, l = 3$ and α is a root of $2 + x^2 + x^3 \in \mathbf{F}_3[x]$.

8.26 Determine the parameters of the QR codes generated by $g_Q(x)$ ($g_N(x)$, respectively) of Exercise 8.25.

   Problems 8.27–8.31 are designed to determine the square root bound on the minimum distance of binary QR codes.

8.27 Let p be a prime of the form $8m \pm 1$. Define

$$E_Q(x) := \begin{cases} 1 + \sum_{i \in N_p} x^i & \text{if } p \text{ is of the form } 8m - 1 \\ \sum_{i \in N_p} x^i & \text{if } p \text{ is of the form } 8m + 1. \end{cases}$$

   (i) Show that θ in Definition 8.3.8 can be chosen properly so that $E_Q(x)$ is an idempotent of the binary QR code C_Q of length p.
   (ii) Put $E_Q(x) = \sum_{i=0}^{p-1} e_i x^i$ and define the $p \times p$ circulant matrix over $\mathbf{F}_2$:

$$G_1 = \begin{pmatrix} e_0 & e_1 & \cdots & e_{p-1} \\ e_{p-1} & e_0 & \cdots & e_{p-2} \\ \cdots & \cdots & \ddots & \cdots \\ e_1 & e_2 & \cdots & e_0 \end{pmatrix}.$$

   Show that every codeword of the binary QR code C_Q of length p is a linear combination of the rows of the matrix

$$G := \begin{pmatrix} 1 \\ G_1 \end{pmatrix}.$$

8.28 Let $u_i \in \mathbf{F}_p$ be the multiplicative inverse of $i \in \mathbf{F}_p^*$. Show that, for any codeword $c(x) = \sum_{i=0}^{p-1} c_i x^i$ of even weight in the binary QR code C_Q of length p, the word $\sum_{i=1}^{p-1} c_i x^{-u_i}$ belongs to C_Q. (Hint: Show that $c(x)$ is a linear combination of the rows of G_1, and then prove that the statement is true if $c(x)$ is a row of G_1.)

8.29 Use Exercise 8.28 to show that the minimum distance of a binary QR code is odd.

8.30 Show that the minimum distance of a binary QR code of length p is at least $\sqrt{p}$.

8.31 Let p be a prime of the form $4k - 1$.
   (i) Show that -1 is a quadratic nonresidue modulo p.

(ii) Show that the minimum distance d of the binary $[p, (p+1)/2]$-QR codes C_Q, C_N satisfies $d^2 - d + 1 \geq p$.

8.32 Let $g_Q(x)$ and $g_N(x)$ be the two polynomials defined in Definition 8.3.8. The binary codes $\tilde{C}_Q$ and $\tilde{C}_N$ generated by $(x-1)g_Q(x)$ and $(x-1)g_N(x)$, respectively, are called *expurgated* QR codes.

(a) Show that $\tilde{C}_Q$ and $\tilde{C}_N$ have dimension $(p-1)/2$ and minimum distance at least $\sqrt{p}$.

(b) If p is of the form $4k-1$, show that $C_Q^\perp = \tilde{C}_Q$ and $C_N^\perp = \tilde{C}_N$.

(c) If p is of the form $4k+1$, show that $C_N^\perp = \tilde{C}_Q$ and $C_Q^\perp = \tilde{C}_N$.

9 Goppa codes

V. D. Goppa described an interesting new class of linear error-correcting codes, commonly called Goppa codes, in the early 1970s. This class of codes includes the narrow-sense BCH codes. It turned out that Goppa codes also form arguably the most interesting subclass of alternant codes, introduced by H. J. Helgert in 1974. The class of alternant codes is a large and interesting family which contains well known codes such as the BCH codes and the Goppa codes.

9.1 Generalized Reed–Solomon codes

We encountered Reed–Solomon (RS) codes in Section 8.2 as a special class of BCH codes. Recall that an RS code over $\mathbf{F}_q$ is a BCH code over $\mathbf{F}_q$ of length $q - 1$ generated by

$$g(x) = (x - \alpha^a)(x - \alpha^{a+1}) \cdots (x - \alpha^{a+\delta-2}),$$

with $a \geq 1$ and $q - 1 \geq \delta \geq 2$, where α is a primitive element of $\mathbf{F}_q$. It is an MDS code with parameters $[q - 1, q - \delta, \delta]$ (cf. Theorem 8.2.3).

Consider the case of the narrow-sense RS codes, i.e., where $a = 1$. In this case, there is an alternative description of the RS code that is convenient for our purpose in this chapter.

Theorem 9.1.1 *Let α be a primitive element of the finite field $\mathbf{F}_q$, and let $q - 1 \geq \delta \geq 2$. The narrow-sense q-ary RS code with generator polynomial*

$$g(x) = (x - \alpha)(x - \alpha^2) \cdots (x - \alpha^{\delta-1})$$

is equal to

$$\{(f(1), f(\alpha), f(\alpha^2), \ldots, f(\alpha^{q-2})) : f(x) \in \mathbf{F}_q[x] \text{ and } \deg(f(x)) < q - \delta\}. \tag{9.1}$$

Proof. It is easy to verify that the set in (9.1) is a vector space over $\mathbf{F}_q$. We first show that it is contained in the RS code generated by $g(x)$.

The codeword $\mathbf{c} = (f(1), f(\alpha), f(\alpha^2), \ldots, f(\alpha^{q-2}))$ corresponds to the polynomial $c(x) = \sum_{i=0}^{q-2} f(\alpha^i) x^i \in \mathbf{F}_q[x]/(x^n - 1)$. We need to show that $g(x)$ divides $c(x)$ (cf. Lemma 8.1.16); i.e.,

$$c(\alpha) = c(\alpha^2) = \ldots = c(\alpha^{\delta-1}) = 0.$$

Note that, for $1 \leq k \leq q - 2$, we have $\sum_{i=0}^{q-2} \alpha^{ik} = ((\alpha^k)^{q-1} - 1)/(\alpha^k - 1) = 0$.

Write $f(x) = \sum_{j=0}^{q-\delta-1} f_j x^j$. Then, for $1 \leq \ell \leq \delta - 1$,

$$c(\alpha^\ell) = \sum_{i=0}^{q-2} f(\alpha^i)(\alpha^\ell)^i = \sum_{i=0}^{q-2} \left(\sum_{j=0}^{q-\delta-1} f_j \alpha^{ij} \right) \alpha^{i\ell} = \sum_{j=0}^{q-\delta-1} f_j \left(\sum_{i=0}^{q-2} \alpha^{i(j+\ell)} \right) = 0,$$

since $1 \leq j + \ell \leq q - 2$.

The map $f \mapsto (f(1), f(\alpha), f(\alpha^2), \ldots, f(\alpha^{q-2}))$ from the set of polynomials in $\mathbf{F}_q[x]$ of degree $< q - \delta$ to the set in (9.1) is injective. (Any $f(x)$ in the kernel of this map must have at least $q - 1 > q - \delta > \deg(f(x))$ zeros, but this is only possible if $f(x)$ is identically equal to 0.) This map is clearly surjective, hence it is an isomorphism of $\mathbf{F}_q$-vector spaces. Therefore, the dimension over $\mathbf{F}_q$ of the vector space in (9.1) is $q - \delta$, which is the dimension of the RS code generated by $g(x)$. Hence, the theorem follows. $\qquad\square$

The following corollary gives another explicit generator matrix for the narrow-sense RS code.

Corollary 9.1.2 *Let α be a primitive element of $\mathbf{F}_q$, and let $q - 1 \geq \delta \geq 2$. The matrix*

$$\begin{pmatrix} 1 & 1 & 1 & \cdots & 1 \\ 1 & \alpha & \alpha^2 & \cdots & \alpha^{q-2} \\ 1 & \alpha^2 & \alpha^4 & \cdots & \alpha^{2(q-2)} \\ \vdots & & \vdots & & \vdots \\ 1 & \alpha^{q-\delta-1} & \alpha^{2(q-\delta-1)} & \cdots & \alpha^{(q-\delta-1)(q-2)} \end{pmatrix}$$

is a generator matrix for the RS code generated by the polynomial

$$g(x) = (x - \alpha)(x - \alpha^2) \cdots (x - \alpha^{\delta-1}).$$

An easy generalization of the description of the RS code in Theorem 9.1.1 leads to a more general class of codes which are also MDS.

Definition 9.1.3 Let $n \leq q$. Let $\alpha = (\alpha_1, \alpha_2, \ldots, \alpha_n)$, where α_i $(1 \leq i \leq n)$ are distinct elements of $\mathbf{F}_q$. Let $\mathbf{v} = (v_1, v_2, \ldots, v_n)$, where $v_i \in \mathbf{F}_q^*$ for all $1 \leq i \leq n$. For $k \leq n$, the *generalized Reed–Solomon code* $GRS_k(\alpha, \mathbf{v})$ is defined to be

$$\{(v_1 f(\alpha_1), v_2 f(\alpha_2), \ldots, v_n f(\alpha_n)) \; : \; f(x) \in \mathbf{F}_q[x] \text{ and } \deg(f(x)) < k\}.$$

The elements $\alpha_1, \alpha_2, \ldots, \alpha_n$ are called the *code locators* of $GRS_k(\alpha, \mathbf{v})$.

Theorem 9.1.4 *The generalized RS code $GRS_k(\alpha, \mathbf{v})$ has parameters $[n, k, n - k + 1]$, so it is an MDS code.*

Proof. It is obvious that $GRS_k(\alpha, \mathbf{v})$ has length n. The same argument as in the proof of Theorem 9.1.1 also shows that its dimension is k. It remains to show that its minimum distance is $n - k + 1$.

To do this, we count the maximum number of zeros in a nonzero codeword. Suppose $f(x)$ is not identically zero. Since $\deg(f(x)) < k$, the polynomial $f(x)$ can only have at most $k - 1$ zeros; i.e., the codeword $(v_1 f(\alpha_1), v_2 f(\alpha_2), \ldots, v_n f(\alpha_n))$ has at most $k - 1$ zeros among its coordinates. In other words, its weight is at least $n - k + 1$, so the minimum distance d of $GRS_k(\alpha, \mathbf{v})$ satisfies $d \geq n - k + 1$. However, the Singleton bound shows that $d \leq n - k + 1$, so $d = n - k + 1$. Hence, $GRS_k(\alpha, \mathbf{v})$ is MDS. $\square$

Remark 9.1.5 In the case where $\mathbf{v} = (1, 1, \ldots, 1)$ and $n < q - 1$, the generalized RS code constructed is often called a *punctured* RS code, as it can be obtained by puncturing an RS code at suitable coordinates.

As for RS codes (cf. Exercise 8.19), the dual of a generalized Reed–Solomon code is again a generalized Reed–Solomon code.

Theorem 9.1.6 *The dual of the generalized Reed–Solomon code $GRS_k(\alpha, \mathbf{v})$ over $\mathbf{F}_q$ of length n is $GRS_{n-k}(\alpha, \mathbf{v}')$ for some $\mathbf{v}' \in (\mathbf{F}_q^*)^n$.*

Proof. First, let $k = n - 1$. From Theorems 5.4.5 and 9.1.4, the dual of $GRS_{n-1}(\alpha, \mathbf{v})$ is an MDS code of dimension 1, so it has parameters $[n, 1, n]$. In particular, its basis consists of a vector $\mathbf{v}' = (v_1', \ldots, v_n')$, where $v_i' \in \mathbf{F}_q^*$ for all $1 \leq i \leq n$. Clearly, this dual code is $GRS_1(\alpha, \mathbf{v}')$.

It follows, in particular, that, for all $f(x) \in \mathbf{F}_q[x]$ of degree $< n - 1$, we have

$$v_1 v_1' f(\alpha_1) + \cdots + v_n v_n' f(\alpha_n) = 0, \tag{9.2}$$

where $\mathbf{v} = (v_1, \ldots, v_n)$.

Now, for arbitrary k, we claim that $GRS_k(\boldsymbol{\alpha}, \mathbf{v})^{\perp} = GRS_{n-k}(\boldsymbol{\alpha}, \mathbf{v}')$.

A typical codeword in $GRS_k(\boldsymbol{\alpha}, \mathbf{v})$ is $(v_1 f(\alpha_1), \ldots, v_n f(\alpha_n))$, where $f(x) \in \mathbf{F}_q[x]$ with degree $\leq k - 1$, while a typical codeword in $GRS_{n-k}(\boldsymbol{\alpha}, \mathbf{v}')$ has the form $(v_1' g(\alpha_1), \ldots, v_n' g(\alpha_n))$, with $g(x) \in \mathbf{F}_q[x]$ of degree $\leq n - k - 1$. Since $\deg(f(x)g(x)) \leq n - 2 < n - 1$, we have

$$(v_1 f(\alpha_1), \ldots, v_n f(\alpha_n)) \cdot (v_1' g(\alpha_1), \ldots, v_n' g(\alpha_n)) = 0$$

from (9.2).

Therefore, $GRS_{n-k}(\boldsymbol{\alpha}, \mathbf{v}') \subseteq GRS_k(\boldsymbol{\alpha}, \mathbf{v})^{\perp}$. Comparing the dimensions of both codes, the theorem follows. $\qquad\square$

Corollary 9.1.7 *A parity-check matrix of $GRS_k(\boldsymbol{\alpha}, \mathbf{v})$ is*

$$
\begin{pmatrix}
v_1' & v_2' & \cdots & v_n' \\
v_1'\alpha_1 & v_2'\alpha_2 & \cdots & v_n'\alpha_n \\
v_1'\alpha_1^2 & v_2'\alpha_2^2 & \cdots & v_n'\alpha_n^2 \\
\vdots & \vdots & & \vdots \\
v_1'\alpha_1^{n-k-1} & v_2'\alpha_2^{n-k-1} & \cdots & v_n'\alpha_n^{n-k-1}
\end{pmatrix}
$$

$$
= \begin{pmatrix}
1 & 1 & \cdots & 1 \\
\alpha_1 & \alpha_2 & \cdots & \alpha_n \\
\alpha_1^2 & \alpha_2^2 & \cdots & \alpha_n^2 \\
\vdots & \vdots & \vdots \\
\alpha_1^{n-k-1} & \alpha_2^{n-k-1} & \cdots & \alpha_n^{n-k-1}
\end{pmatrix}
\begin{pmatrix}
v_1' & 0 & \cdots & 0 \\
0 & v_2' & \cdots & 0 \\
& & \ddots & \\
\vdots & \vdots & & \vdots \\
0 & 0 & \cdots & v_n'
\end{pmatrix}.
$$

Remark 9.1.8 Recall that $\mathbf{v}' = (v_1', \ldots, v_n')$ is any vector that generates the dual of $GRS_{n-1}(\boldsymbol{\alpha}, \mathbf{v})$, so it is not unique (cf. Exercise 9.1). In particular, the parity-check matrix in Corollary 9.1.7 is also not unique.

9.2 Alternant codes

An interesting family of codes arising from the generalized RS codes of the previous section is the class of alternant codes. This is quite a large family that includes the Hamming codes and the BCH codes.

We use the same notation as in the previous section, except that the generalized RS codes are now defined over $\mathbf{F}_{q^m}$, for some $m \geq 1$.

Definition 9.2.1 An *alternant* code $A_k(\boldsymbol{\alpha}, \mathbf{v}')$ over the finite field $\mathbf{F}_q$ is the subfield subcode $GRS_k(\boldsymbol{\alpha}, \mathbf{v})|_{\mathbf{F}_q}$, where $GRS_k(\boldsymbol{\alpha}, \mathbf{v})$ is a generalized RS code over $\mathbf{F}_{q^m}$, for some $m \geq 1$.

Remark 9.2.3 below explains why we have chosen $\mathbf{v}'$ in the notation for the alternant code instead of $\mathbf{v}$.

Proposition 9.2.2 *The alternant code* $\mathcal{A}_k(\boldsymbol{\alpha}, \mathbf{v}')$ *has parameters* $[n, k', d]$, *where* $mk - (m-1)n \le k' \le k$ *and* $d \ge n - k + 1$.

Proof. By Theorem 9.1.4, $GRS_k(\boldsymbol{\alpha}, \mathbf{v})$ has parameters $[n, k, n-k+1]$. Hence, $\mathcal{A}_k(\boldsymbol{\alpha}, \mathbf{v}')$ clearly has length n, and its dimension k' trivially satisfies $k' \le k$. The result follows from Theorem 6.3.5. $\qquad\square$

Remark 9.2.3 It follows directly from Definition 9.2.1 and Corollary 9.1.7 that $\mathcal{A}_k(\boldsymbol{\alpha}, \mathbf{v}')$ is none other than

$$\{\mathbf{c} \in \mathbf{F}_q^n \ : \ \mathbf{c}H^{\mathrm{T}} = \mathbf{0}\},$$

where H is the matrix in Corollary 9.1.7. Since H is determined by $\boldsymbol{\alpha}$ and $\mathbf{v}'$, it is appropriate for the notation for the alternant code to be expressed in terms of $\boldsymbol{\alpha}$ and $\mathbf{v}'$.

Recall that every element $\beta \in \mathbf{F}_{q^m}$ can be written uniquely in the form $\sum_{i=0}^{m-1} \beta_i \alpha^i$, where α is a primitive element of $\mathbf{F}_{q^m}$ and $\beta_i \in \mathbf{F}_q$, for all $0 \le i \le m - 1$. Therefore, if we replace every entry β of H by the column vector $(\beta_0, \ldots, \beta_{m-1})^{\mathrm{T}}$, we obtain an $(n - k)m \times n$ matrix $\overline{H}$ with entries in $\mathbf{F}_q$ such that $\mathcal{A}_k(\boldsymbol{\alpha}, \mathbf{v}')$ is

$$\{\mathbf{c} \in \mathbf{F}_q^n \ : \ \mathbf{c}\overline{H}^{\mathrm{T}} = \mathbf{0}\}.$$

This matrix $\overline{H}$ plays the role of a parity-check matrix of $\mathcal{A}_k(\boldsymbol{\alpha}, \mathbf{v}')$, except that its rows are not necessarily linearly independent, so we refrain from calling it a parity-check matrix of $\mathcal{A}_k(\boldsymbol{\alpha}, \mathbf{v}')$. (However, this appellation is used in some books.)

We now look at some examples of alternant codes.

Example 9.2.4 (i) Let $q = 2$ and let m be any integer ≥ 3. Let α be a primitive element of $\mathbf{F}_{2^m}$. Set

$$\mathbf{v}' = (1, \alpha, \alpha^2, \ldots, \alpha^{2^m - 2}).$$

For any $\boldsymbol{\alpha} = (\alpha_1, \ldots, \alpha_{2^m - 1})$, where $\{\alpha_1, \ldots, \alpha_{2^m-1}\} = \mathbf{F}_{2^m}^*$, the alternant code $\mathcal{A}_{2^m - 2}(\boldsymbol{\alpha}, \mathbf{v}')$ is

$$\mathcal{A}_{2^m - 2}(\boldsymbol{\alpha}, \mathbf{v}') = \left\{\mathbf{c} \in \mathbf{F}_2^{2^m - 1} \ : \ \mathbf{c}(1, \alpha, \alpha^2, \ldots, \alpha^{2^m - 2})^{\mathrm{T}} = \mathbf{0}\right\}.$$

It is clear that, for $H = (1, \alpha, \alpha^2, \ldots, \alpha^{2^m-2})$, $\overline{H}$ is an $m \times (2^m - 1)$ matrix whose columns are all the nonzero vectors in $\mathbf{F}_2^m$. Recall that this is a parity-check matrix for the binary Hamming code Ham$(m, 2)$, so $\mathcal{A}_{2^m-2}(\alpha, \mathbf{v}') =$ Ham$(m, 2)$.

(ii) For any q and m, recall from (8.1) that a BCH code over $\mathbf{F}_q$ is a code consisting of all $\mathbf{c} \in \mathbf{F}_q^n$ that satisfy $\mathbf{c}H'^{\mathrm{T}} = \mathbf{0}$, where

$$H' = \begin{pmatrix} 1 & \alpha^a & \alpha^{2a} & \cdots & \alpha^{a(n-1)} \\ 1 & \alpha^{a+1} & \alpha^{2(a+1)} & \cdots & \alpha^{(a+1)(n-1)} \\ 1 & \alpha^{a+2} & \alpha^{2(a+2)} & \cdots & \alpha^{(a+2)(n-1)} \\ \vdots & \vdots & & \ddots & \vdots \\ 1 & \alpha^{a+\delta-2} & \alpha^{2(a+\delta-2)} & \cdots & \alpha^{(a+\delta-2)(n-1)} \end{pmatrix}$$

$$= \begin{pmatrix} 1 & 1 & 1 & \cdots & 1 \\ 1 & \alpha & \alpha^2 & \cdots & \alpha^{n-1} \\ 1 & \alpha^2 & \alpha^4 & \cdots & \alpha^{2(n-1)} \\ \vdots & \vdots & & \ddots & \vdots \\ 1 & \alpha^{\delta-2} & \alpha^{2(\delta-2)} & \cdots & \alpha^{(\delta-2)(n-1)} \end{pmatrix} \begin{pmatrix} 1 & 0 & \cdots & 0 \\ 0 & \alpha^a & \cdots & 0 \\ & & \ddots & \\ \vdots & \vdots & & \vdots \\ 0 & 0 & \cdots & \alpha^{a(n-1)} \end{pmatrix},$$

which is exactly in the form of Corollary 9.1.7. Therefore, a BCH code is also an alternant code.

(iii) Let $q = 2$ and $m = 3$, and set $n = 6$. Let α be a primitive element of $\mathbf{F}_8$ that satisfies $\alpha^3 + \alpha + 1 = 0$. Take $\mathbf{v}' = (1, \ldots, 1)$ and $\alpha = (\alpha, \alpha^2, \ldots, \alpha^6)$. Then $\mathcal{A}_3(\alpha, \mathbf{v}') = \{\mathbf{c} \in \mathbf{F}_2^6 : \mathbf{c}H^{\mathrm{T}} = \mathbf{0}\}$, where

$$H = \begin{pmatrix} 1 & 1 & 1 & 1 & 1 & 1 \\ \alpha & \alpha^2 & \alpha^3 & \alpha^4 & \alpha^5 & \alpha^6 \\ \alpha^2 & \alpha^4 & \alpha^6 & \alpha & \alpha^3 & \alpha^5 \end{pmatrix}.$$

Then

$$\overline{H} = \begin{pmatrix} 1 & 1 & 1 & 1 & 1 & 1 \\ 0 & 0 & 0 & 0 & 0 & 0 \\ 0 & 0 & 0 & 0 & 0 & 0 \\ 0 & 0 & 1 & 0 & 1 & 1 \\ 1 & 0 & 1 & 1 & 1 & 0 \\ 0 & 1 & 0 & 1 & 1 & 1 \\ 0 & 0 & 1 & 0 & 1 & 1 \\ 0 & 1 & 0 & 1 & 1 & 1 \\ 1 & 1 & 1 & 0 & 0 & 1 \end{pmatrix},$$

which has the following reduced row echelon form:

$$\begin{pmatrix} 1 & 0 & 0 & 0 & 1 & 1 \\ 0 & 1 & 0 & 0 & 0 & 1 \\ 0 & 0 & 1 & 0 & 1 & 1 \\ 0 & 0 & 0 & 1 & 1 & 0 \\ 0 & 0 & 0 & 0 & 0 & 0 \\ 0 & 0 & 0 & 0 & 0 & 0 \\ 0 & 0 & 0 & 0 & 0 & 0 \\ 0 & 0 & 0 & 0 & 0 & 0 \\ 0 & 0 & 0 & 0 & 0 & 0 \end{pmatrix}.$$

Hence, it follows that $A_3(\alpha, \mathbf{v}')$ has a generator matrix

$$\begin{pmatrix} 1 & 0 & 1 & 1 & 1 & 0 \\ 1 & 1 & 1 & 0 & 0 & 1 \end{pmatrix},$$

so it is a [6, 2, 4]-code.

The following description of the dual of an alternant code is an immediate consequence of Theorems 6.3.9 and 9.1.6.

Theorem 9.2.5 *The dual of the alternant code $A_k(\alpha, \mathbf{v}')$ is*

$$\mathrm{Tr}_{\mathbf{F}_{q^m}/\mathbf{F}_q}(GRS_{n-k}(\alpha, \mathbf{v}')).$$

The following theorem shows the existence of an alternant code with certain parameters.

Theorem 9.2.6 *Given any positive integers n, h, δ, m, there exists an alternant code $A_k(\alpha, \mathbf{v}')$ over $\mathbf{F}_q$, which is the subfield subcode of a generalized RS code over $\mathbf{F}_{q^m}$, with parameters $[n, k', d]$, where $k' \geq h$ and $d \geq \delta$, so long as*

$$\sum_{w=0}^{\delta-1}(q-1)^w \binom{n}{w} < (q^m - 1)^{\lfloor(n-h)/m\rfloor}. \tag{9.3}$$

Proof. For any vector $\mathbf{c} \in \mathbf{F}_q^n$, let

$$R(\alpha, k, \mathbf{c}) = \{\mathbf{v} \in (\mathbf{F}_{q^m}^*)^n \ : \ \mathbf{c} \in GRS_k(\alpha, \mathbf{v})\}.$$

Writing $\mathbf{c} = (c_1, \ldots, c_n)$ and $\mathbf{v} = (v_1, \ldots, v_n)$, we have that $c_i = v_i f(\alpha_i)$, where $f(x) \in \mathbf{F}_q[x]$ has degree $< k$, for all $1 \leq i \leq n$. For a fixed $\mathbf{c}$, $f(x)$ is fixed once k values of v_i are chosen. Therefore,

$$|R(\alpha, k, \mathbf{c})| \leq (q^m - 1)^k.$$

The number of vectors $\mathbf{c} \in \mathbf{F}_q^n$ of weight $< \delta$ is given by $\sum_{w=0}^{\delta-1} \binom{n}{w}(q-1)^w$, so, taking $k = n - \lfloor (n-h)/m \rfloor$, we have

$$\left| \bigcup_{\substack{\mathrm{wt}(\mathbf{c})<\delta \\ \mathbf{c}\in\mathbf{F}_q^n}} R(\alpha, k, \mathbf{c}) \right| \leq \left(\sum_{w=0}^{\delta-1} \binom{n}{w}(q-1)^w \right) (q^m - 1)^{n-\lfloor(n-h)/m\rfloor}.$$

Now,

$$|(\mathbf{F}_{q^m}^*)^n| = (q^m - 1)^n.$$

Therefore, if (9.3) is satisfied, then

$$\bigcup_{\substack{\mathrm{wt}(\mathbf{c})<\delta \\ \mathbf{c}\in\mathbf{F}_q^n}} R(\alpha, k, \mathbf{c})$$

is strictly smaller than $(\mathbf{F}_{q^m}^*)^n$; i.e., there exists $\mathbf{v} \in (\mathbf{F}_{q^m}^*)^n$ such that $GRS_k(\alpha, \mathbf{v})$ does not contain any vector of $\mathbf{F}_q^n$ of weight $< \delta$. Hence, the alternant code $A_k(\alpha, \mathbf{v}')$ has distance $\geq \delta$. Its length is clearly n. Since $k = n - \lfloor(n-h)/m\rfloor \geq ((m-1)n+h)/m$, Proposition 9.2.2 implies that the dimension k' satisfies $k' \geq mk - (m-1)n \geq h$. □

9.3   Goppa codes

One of the most interesting subclasses of alternant codes is the family of Goppa codes, introduced by V. D. Goppa [3, 4] in the early 1970s. This family also contains long codes that have good parameters. Goppa codes are used also in cryptography – the McEliece cryptosystem and the Niederreiter cryptosystem are examples of public-key cryptosystems that use Goppa codes.

Definition 9.3.1 Let $g(z)$ be a polynomial in $\mathbf{F}_{q^m}[z]$ for some fixed m, and let $L = \{\alpha_1, \ldots, \alpha_n\}$ be a subset of $\mathbf{F}_{q^m}$ such that $L \cap \{\text{zeros of } g(z)\} = \emptyset$. For $\mathbf{c} = (c_1, \ldots, c_n) \in \mathbf{F}_q^n$, let

$$R_{\mathbf{c}}(z) = \sum_{i=1}^n \frac{c_i}{z - \alpha_i}.$$

The *Goppa code* $\Gamma(L, g)$ is defined as

$$\Gamma(L, g) = \{\mathbf{c} \in \mathbf{F}_q^n : R_{\mathbf{c}}(z) \equiv 0 \ (\mathrm{mod}\ g(z))\}.$$

The polynomial $g(z)$ is called the *Goppa polynomial*. When $g(z)$ is irreducible, $\Gamma(L, g)$ is called an *irreducible* Goppa code.

Remark 9.3.2 (i) Notice that

$$(z - \alpha_i) \left(-\frac{g(z) - g(\alpha_i)}{z - \alpha_i} g(\alpha_i)^{-1} \right) \equiv 1 \ (\mathrm{mod} \ g(z)),$$

and, since $(g(z) - g(\alpha_i))/(z - \alpha_i)$ is a polynomial, it follows that, modulo $g(z)$, $1/(z - \alpha_i)$ may be regarded as a polynomial; i.e.,

$$\frac{1}{z - \alpha_i} \equiv -\frac{g(z) - g(\alpha_i)}{z - \alpha_i} g(\alpha_i)^{-1} \ (\mathrm{mod} \ g(z)).$$

Hence, the congruence $R_{\mathbf{c}}(z) \equiv 0 \ (\mathrm{mod} \ g(z))$ in the definition of $\Gamma(L, g)$ means that $g(z)$ divides the polynomial

$$\sum_{i=1}^{n} c_i \frac{g(z) - g(\alpha_i)}{z - \alpha_i} g(\alpha_i)^{-1}.$$

However, noting that $(g(z) - g(\alpha_i))/(z - \alpha_i)$ is a polynomial of degree $< t$ if $g(z)$ has degree t, it follows that $\mathbf{c} \in \Gamma(L, g)$ if and only if

$$\sum_{i=1}^{n} c_i \frac{g(z) - g(\alpha_i)}{z - \alpha_i} g(\alpha_i)^{-1} = 0 \tag{9.4}$$

as a polynomial.

(ii) It is clear from the definition that Goppa codes are linear.

The next proposition shows immediately that Goppa codes are examples of alternant codes.

Proposition 9.3.3 *For a given Goppa polynomial $g(z)$ of degree t and $L = \{\alpha_1, \ldots, \alpha_n\}$, we have $\Gamma(L, g) = \{\mathbf{c} \in \mathbf{F}_q^n : \mathbf{c}H^{\mathrm{T}} = \mathbf{0}\}$, where*

$$H = \begin{pmatrix} g(\alpha_1)^{-1} & \cdots & g(\alpha_n)^{-1} \\ \alpha_1 g(\alpha_1)^{-1} & \cdots & \alpha_n g(\alpha_n)^{-1} \\ \vdots & \cdots & \vdots \\ \alpha_1^{t-1} g(\alpha_1)^{-1} & \cdots & \alpha_n^{t-1} g(\alpha_n)^{-1} \end{pmatrix}.$$

Proof. Recall that $\mathbf{c} \in \Gamma(L, g)$ if and only if (9.4) holds.

Substituting $g(z) = \sum_{i=0}^{t} g_i z^i$ into (9.4), and equating the coefficients of the various powers of z to 0, it follows that $\mathbf{c} \in \Gamma(L, g)$ if and only if $\mathbf{c}H'^{\mathrm{T}} = \mathbf{0}$, where

$$H' = \begin{pmatrix} g_t g(\alpha_1)^{-1} & \cdots & g_t g(\alpha_n)^{-1} \\ (g_{t-1} + \alpha_1 g_t) g(\alpha_1)^{-1} & \cdots & (g_{t-1} + \alpha_n g_t) g(\alpha_n)^{-1} \\ \vdots & \cdots & \vdots \\ (g_1 + \alpha_1 g_2 + \cdots + \alpha_1^{t-1} g_t) g(\alpha_1)^{-1} & \cdots & (g_1 + \alpha_n g_2 + \cdots + \alpha_n^{t-1} g_t) g(\alpha_n)^{-1} \end{pmatrix}.$$

We see easily that H' can also be decomposed as

$$
\begin{pmatrix}
g_t & 0 & \cdots & 0 \\
g_{t-1} & g_t & \cdots & 0 \\
g_{t-2} & g_{t-1} & \cdots & 0 \\
\vdots & & \ddots & \vdots \\
g_1 & g_2 & \cdots & g_t
\end{pmatrix}
\begin{pmatrix}
1 & 1 & \cdots & 1 \\
\alpha_1 & \alpha_2 & \cdots & \alpha_n \\
\alpha_1^2 & \alpha_2^2 & \cdots & \alpha_n^2 \\
\vdots & & \ddots & \vdots \\
\alpha_1^{t-1} & \alpha_2^{t-1} & \cdots & \alpha_n^{t-1}
\end{pmatrix}
\begin{pmatrix}
v_1' & 0 & \cdots & 0 \\
0 & v_2' & \cdots & 0 \\
& & \ddots & \\
0 & 0 & \cdots & v_n'
\end{pmatrix},
$$

where $v_i' = g(\alpha_i)^{-1}$ for $1 \le i \le n$.

It now follows from Exercise 4.38 that $\mathbf{c} \in \Gamma(L, g)$ if and only if $\mathbf{c}H^{\mathrm{T}} = \mathbf{0}$, where

$$
H =
\begin{pmatrix}
1 & 1 & \cdots & 1 \\
\alpha_1 & \alpha_2 & \cdots & \alpha_n \\
\alpha_1^2 & \alpha_2^2 & \cdots & \alpha_n^2 \\
\vdots & & \ddots & \vdots \\
\alpha_1^{t-1} & \alpha_2^{t-1} & \cdots & \alpha_n^{t-1}
\end{pmatrix}
\begin{pmatrix}
g(\alpha_1)^{-1} & 0 & \cdots & 0 \\
0 & g(\alpha_2)^{-1} & \cdots & 0 \\
& & \ddots & \\
\vdots & \vdots & & \vdots \\
0 & 0 & \cdots & g(\alpha_n)^{-1}
\end{pmatrix}.
$$

$\square$

Corollary 9.3.4 *For a given Goppa polynomial $g(z)$ of degree t and $L = \{\alpha_1, \ldots, \alpha_n\}$, the Goppa code $\Gamma(L, g)$ is the alternant code $\mathcal{A}_{n-t}(\alpha, \mathbf{v}')$, where $\alpha = (\alpha_1, \ldots, \alpha_n)$ and $\mathbf{v}' = (g(\alpha_1)^{-1}, \ldots, g(\alpha_n)^{-1})$.*

We can also obtain directly a description of the Goppa code as a subfield subcode of a generalized RS code.

Theorem 9.3.5 *With notation as above, the Goppa code $\Gamma(L, g)$ is $GRS_{n-t}(\alpha, \mathbf{v})|_{\mathbf{F}_q}$, where $\mathbf{v} = (v_1, \ldots, v_n)$ with $v_i = g(\alpha_i)/(\prod_{j \ne i}(\alpha_i - \alpha_j))$, for all $1 \le i \le n$.*

Proof. From Proposition 9.3.3, it is clear that $\Gamma(L, g) = GRS_t(\alpha, \mathbf{v}')^{\perp}|_{\mathbf{F}_q}$, where $\mathbf{v}' = (g(\alpha_1)^{-1}, \ldots, g(\alpha_n)^{-1})$. Hence, it is enough to show that $GRS_t(\alpha, \mathbf{v}')^{\perp} = GRS_{n-t}(\alpha, \mathbf{v})$ (cf. Theorem 9.1.6); i.e.,

$$
v_1 g(\alpha_1)^{-1} f(\alpha_1) + \cdots + v_n g(\alpha_n)^{-1} f(\alpha_n) = 0,
$$

where $\mathbf{v} = (v_1, \ldots, v_n)$ with $v_i = g(\alpha_i)/(\prod_{j \ne i}(\alpha_i - \alpha_j))$, for all $1 \le i \le n$, and for all polynomials $f(x) \in \mathbf{F}_{q^m}[x]$ of degree $\le n - 2$.

Since $f(x)$ is a polynomial of degree $\le n - 2$, it is determined by its values at $\le n - 1$ points, so it follows that (cf. Exercise 3.26)

$$
f(z) = \sum_{i=1}^{n} f(\alpha_i) \left(\prod_{j \ne i} \frac{z - \alpha_j}{\alpha_i - \alpha_j} \right).
$$

Equating the coefficients of z^{n-1}, we obtain (since $\deg(f(x)) \leq n - 2$)

$$0 = \sum_{i=1}^{n} \frac{f(\alpha_i)}{\prod_{j \neq i}(\alpha_i - \alpha_j)} = v_1 g(\alpha_1)^{-1} f(\alpha_1) + \cdots + v_n g(\alpha_n)^{-1} f(\alpha_n).$$

$\square$

By Proposition 9.2.2, Corollary 9.3.4 (or, equivalently, Theorem 9.3.5) also gives immediately a bound for both the dimension and the minimum distance of a Goppa code.

Corollary 9.3.6 *For a given Goppa polynomial $g(z)$ of degree t and $L = \{\alpha_1, \ldots, \alpha_n\}$, the Goppa code $\Gamma(L, g)$ is a linear code over $\mathbf{F}_q$ with parameters $[n, k, d]$, where $k \geq n - mt$ and $d \geq t + 1$.*

The following description of the dual of a Goppa code now follows immediately from Theorem 9.2.5.

Corollary 9.3.7 *With notation as above, the dual of the Goppa code $\Gamma(L, g)$ is the trace code $\mathrm{Tr}_{\mathbf{F}_{q^m}/\mathbf{F}_q}(GRS_t(\alpha, \mathbf{v}'))$, where $\mathbf{v}' = (g(\alpha_1)^{-1}, \ldots, g(\alpha_n)^{-1})$.*

When $q = 2$, i.e., in the binary case, a sharpening of the lower bound on d can be obtained.

For a given polynomial $g(z)$, we write $\tilde{g}(z)$ for the lowest degree perfect square polynomial that is divisible by $g(z)$. Denote by $\tilde{t}$ the degree of $\tilde{g}(z)$.

For a vector $\mathbf{c} = (c_1, \ldots, c_n) \in \mathbf{F}_q^n$ of weight w, with $c_{i_1} = \cdots = c_{i_w} = 1$, say, let

$$f_{\mathbf{c}}(z) = \prod_{j=1}^{w} (z - \alpha_{i_j}).$$

Taking its derivative yields

$$f_{\mathbf{c}}'(z) = \sum_{\ell=1}^{w} \prod_{j \neq \ell} (z - \alpha_{i_j}).$$

Hence, we have

$$R_{\mathbf{c}}(z) = \frac{f_{\mathbf{c}}'(z)}{f_{\mathbf{c}}(z)}. \tag{9.5}$$

Proposition 9.3.8 *Let $q = 2$. With notation as above, $\mathbf{c} \in \mathbf{F}_2^n$ belongs to $\Gamma(L, g)$ if and only if $\tilde{g}(z)$ divides $f_{\mathbf{c}}'(z)$. Consequently, the minimum distance d of $\Gamma(L, g)$ satisfies $d \geq \tilde{t} + 1$. In particular, if $g(z)$ has no multiple root (i.e., $g(z)$ is a separable polynomial), then $d \geq 2t + 1$.*

Proof. By definition, $\mathbf{c} \in \Gamma(L, g)$ if and only if $R_{\mathbf{c}}(z) \equiv 0 \pmod{g(z)}$. From (9.5), and noting that $f_{\mathbf{c}}(z)$ and $g(z)$ have no common factors, it follows that $\mathbf{c} \in \Gamma(L, g)$ if and only if $g(z)$ divides $f'_{\mathbf{c}}(z)$. However, as we are working in characteristic 2, $f'_{\mathbf{c}}(z)$, being the derivative of a polynomial, contains only even powers of z and is hence a perfect square polynomial. Therefore, $g(z)$ divides $f'_{\mathbf{c}}(z)$ if and only if $\tilde{g}(z)$ divides $f'_{\mathbf{c}}(z)$. This proves the first statement of the proposition.

If $\mathbf{c}$ is a codeword of minimum weight d in $\Gamma(L, g)$, then $f_{\mathbf{c}}(z)$ has degree d, so $f'_{\mathbf{c}}(z)$ has degree $\leq d - 1$. The condition that $\tilde{g}(z)$ divides $f'_{\mathbf{c}}(z)$ implies that $d - 1 \geq \deg(f'_{\mathbf{c}}(z)) \geq \deg(\tilde{g}(z)) = \tilde{t}$.

If $g(z)$ has no multiple root, then clearly $\tilde{g}(z) = (g(z))^2$, so $\tilde{t} = 2t$.   □

Remark 9.3.9 (i) When $g(z)$ is separable, the Goppa code $\Gamma(L, g)$ is said to be *separable*.

(ii) If it is known that the minimum distance d of $\Gamma(L, g)$ is even, then the bounds above can be slightly improved to $d \geq \tilde{t} + 2$ and $d \geq 2t + 2$.

Example 9.3.10 (i) Let $q = 2$, let $g(z) = z$ and set $L = \mathbf{F}^*_{2^m}$. The Goppa code $\Gamma(L, g)$ is then $\{\mathbf{c} \in \mathbf{F}_2^{2^m - 1} : \mathbf{c}H^{\mathrm{T}} = \mathbf{0}\}$, where

$$H = (1, \alpha, \alpha^2, \ldots, \alpha^{2^m - 2}),$$

with α a primitive element of $\mathbf{F}_{2^m}$. As we have seen in Example 9.2.4(i), this is none other than the binary Hamming code Ham$(m, 2)$.

(ii) For any q, take $g(z) = z^t$ and let $L = \{1, \alpha^{-1}, \alpha^{-2}, \ldots, \alpha^{-(q^m - 2)}\}$, where α is a primitive element of $\mathbf{F}_{q^m}$. (Hence, $n = q^m - 1$.) Then $\Gamma(L, g) = \{\mathbf{c} \in \mathbf{F}_q^n : \mathbf{c}H^{\mathrm{T}} = \mathbf{0}\}$, where

$$H = \begin{pmatrix} 1 & \alpha^t & \alpha^{2t} & \cdots & \alpha^{(n-1)t} \\ 1 & \alpha^{t-1} & \alpha^{2(t-1)} & \cdots & \alpha^{(n-1)(t-1)} \\ \vdots & \vdots & \vdots & \ddots & \vdots \\ 1 & \alpha & \alpha^2 & \cdots & \alpha^{n-1} \end{pmatrix}.$$

Comparing with (8.1), we see that $\Gamma(L, g)$ is precisely a narrow-sense BCH code.

(iii) Let $q = 2$ and take $g(z) = \alpha^3 + z + z^2$, where α is a primitive element of $\mathbf{F}_8$ that satisfies $\alpha^3 + \alpha + 1 = 0$. Let $L = \mathbf{F}_8$, so $n = 8$ and $m = 3$. Then $\Gamma(L, g) = \{\mathbf{c} \in \mathbf{F}_2^8 : \mathbf{c}H^{\mathrm{T}} = \mathbf{0}\}$, where

$$H = \begin{pmatrix} \alpha^4 & \alpha^4 & \alpha & 1 & \alpha & \alpha^2 & \alpha^2 & 1 \\ 0 & \alpha^4 & \alpha^2 & \alpha^2 & \alpha^4 & \alpha^6 & 1 & \alpha^6 \end{pmatrix}.$$

Replacing each entry in H by a column vector in $\mathbf{F}_2^3$, we obtain a matrix $\overline{H}$ which has the following reduced row echelon form:

$$\begin{pmatrix} 1 & 0 & 0 & 0 & 0 & 0 & 1 & 1 \\ 0 & 1 & 0 & 0 & 0 & 0 & 1 & 0 \\ 0 & 0 & 1 & 0 & 0 & 0 & 1 & 1 \\ 0 & 0 & 0 & 1 & 0 & 0 & 0 & 1 \\ 0 & 0 & 0 & 0 & 1 & 0 & 1 & 0 \\ 0 & 0 & 0 & 0 & 0 & 1 & 1 & 1 \end{pmatrix},$$

so $\Gamma(L, g)$ has a generator matrix

$$\begin{pmatrix} 1 & 1 & 1 & 0 & 1 & 1 & 1 & 0 \\ 1 & 0 & 1 & 1 & 0 & 1 & 0 & 1 \end{pmatrix}.$$

Therefore, $\Gamma(L, g)$ has parameters $[8, 2, 5]$.

In both Examples 9.3.10(i) and (iii), the bound $d \geq 2t + 1$ in Proposition 9.3.8 is attained. The following theorem shows the existence of a Goppa code of certain parameters.

Theorem 9.3.11 *There is a q-ary Goppa code $\Gamma(L, g)$, where $g(z)$ is an irreducible polynomial in $\mathbf{F}_{q^m}[z]$ of degree t and $L = \mathbf{F}_{q^m}$, of parameters $[q^m, k, d]$, where $k \geq q^m - mt$, provided*

$$\sum_{w=t+1}^{d-1} \left\lfloor \frac{w-1}{t} \right\rfloor (q-1)^w \binom{q^m}{w} < \frac{1}{t} q^{mt} (1 - (t-1)q^{-mt/2}). \qquad (9.6)$$

Proof. Write $n = q^m$. Let $\mathbf{c} = (c_1, \ldots, c_n) \in \mathbf{F}_q^n$ be of weight w, with $c_{i_1} \neq 0, \ldots, c_{i_w} \neq 0$. Then $\mathbf{c} \in \Gamma(L, g)$ if and only if $R_{\mathbf{c}}(z) \equiv 0 \pmod{g(z)}$. Since $\mathbf{c}$ has weight w, $R_{\mathbf{c}}(z) = h_{\mathbf{c}}(z)/\prod_{j=1}^{w}(z - \alpha_{i_j})$, where $h_{\mathbf{c}}(z)$ has degree $\leq w - 1$ and $\prod_{j=1}^{w}(z - \alpha_{i_j})$ has no common factor with $g(z)$. Therefore, $\mathbf{c} \in \Gamma(L, g)$ if and only if $g(z)$ divides $h_{\mathbf{c}}(z)$. The number of irreducible polynomials $g(z)$ of degree t that can divide a given $h_{\mathbf{c}}(z)$ is at most $\lfloor (w-1)/t \rfloor$, so the number of $\Gamma(L, g)$ containing a given $\mathbf{c}$ of weight w, with $g(z)$ irreducible of degree t, is at most $\lfloor (w-1)/t \rfloor$.

The number of $\mathbf{c}$ of a given weight w is $(q-1)^w \binom{q^m}{w}$, so the total number of $\Gamma(L, g)$ containing at least a word of weight $< d$ is $\leq \sum_{w=t+1}^{d-1}(q-1)^w \binom{q^m}{w} \lfloor (w-1)/t \rfloor$. (Since $g(z)$ has degree t, Corollary 9.3.6 implies that $\Gamma(L, g)$ does not have any nonzero words of degree $\leq t$, so the sum begins with $w = t + 1$.)

The number of irreducible polynomials in $\mathbf{F}_{q^m}[z]$ of degree t is given by $I_{q^m}(t) = \left(\sum_{s|t} \mu(s) q^{mt/s} \right) / t$ (cf. Exercise 3.28), where μ is the Möbius function. For $2 \leq s \leq t$, clearly $\mu(s)q^{mt/s} \geq -q^{mt/2}$. Hence, with $d(t)$ denoting the number of positive divisors of t, we have

$$I_{q^m}(t) = \frac{1}{t} \sum_{s|t} \mu(s) q^{mt/s} \geq \frac{1}{t}(q^{mt} - (d(t)-1)q^{mt/2}) \geq \frac{1}{t}(q^{mt} - (t-1)q^{mt/2}).$$

Therefore, if (9.6) holds, then there is at least one irreducible polynomial $g(z)$ in $\mathbf{F}_{q^m}[z]$ of degree t such that $\Gamma(L, g)$ does not contain any nonzero word of weight $< d$; i.e., the minimum distance of $\Gamma(L, g)$ is at least d. ☐

9.4   Sudan decoding for generalized RS codes

For a linear code C, a *list-decoding* with error-bound τ produces a list of all the codewords $\mathbf{c} \in C$ that are within Hamming distance τ from the received word. Consider the q-ary generalized Reed–Solomon code $GRS_{k+1}(\boldsymbol{\alpha}, \mathbf{1})$, where $\boldsymbol{\alpha} = (\alpha_1, \ldots, \alpha_n)$ with $\alpha_i \in \mathbf{F}_q$, for $1 \leq i \leq n$, and $\mathbf{1} = (1, \ldots, 1)$; i.e.,

$GRS_{k+1}(\boldsymbol{\alpha}, \mathbf{1})$
$$= \{(f(\alpha_1), f(\alpha_2), \ldots, f(\alpha_n)) : f(x) \in \mathbf{F}_q[x] \text{ and } \deg(f(x)) \leq k\}. \quad (9.7)$$

Recall that $GRS_{k+1}(\boldsymbol{\alpha}, \mathbf{1})$ is an $[n, k+1, n-k]$-linear code over $\mathbf{F}_q$.

In this section, we discuss an algorithm, due basically to M. Sudan, for a list-decoding for $GRS_{k+1}(\boldsymbol{\alpha}, \mathbf{1})$. It is one of the most effective decoding schemes currently available for such codes. Modifications of this algorithm are also available for the decoding of some other codes discussed in this chapter, but we restrict our discussion to this generalized RS code. For more details, the reader may refer to refs. [6], [18] and [21].

For $GRS_{k+1}(\boldsymbol{\alpha}, \mathbf{1})$ $(0 < k < n)$ and a received word $(\beta_1, \beta_2, \ldots, \beta_n) \in \mathbf{F}_q^n$, let $\mathcal{P} = \{(\alpha_i, \beta_i) : 1 \leq i \leq n\}$, and let t be a positive integer $< n$.

In general, a list-decoding with error-bound $\tau = n - t$ solves the following polynomial reconstruction problem:

$(\mathcal{P}, k, t)$-**reconstruction** *For $\mathcal{P}, k, t$ as above, reconstruct the set, denoted by $\Omega(\mathcal{P}, k, t)$, of all the polynomials $f(x) \in \mathbf{F}_q[x]$, with $\deg(f(x)) \leq k$, which satisfy*

$$|\{(\alpha, \beta) \in \mathcal{P} : f(\alpha) = \beta\}| \geq t. \quad (9.8)$$

The Sudan algorithm is a polynomial-time list-decoding algorithm for $GRS_{k+1}(\boldsymbol{\alpha}, \mathbf{1})$ that solves the $(\mathcal{P}, k, t)$-reconstruction problem in two stages as follows:

- *Generation of the $(\mathcal{P}, k, t)$-polynomial.* Generate a nonzero bivariate polynomial $Q(x, y) \in \mathbf{F}_q[x, y]$, called the $(\mathcal{P}, k, t)$-*polynomial*, by solving a linear system in polynomial time such that $y - f(x)$ divides $Q(x, y)$, for all $f(x) \in \Omega(\mathcal{P}, k, t)$.
- *Factorization of the $(\mathcal{P}, k, t)$-polynomial.* Factorize the $(\mathcal{P}, k, t)$-polynomial $Q(x, y)$ and then output $\Omega(\mathcal{P}, k, t)$, which is the set of polynomials $f(x) \in \mathbf{F}_q[x]$, with $\deg(f(x)) \le k$, such that $y - f(x)$ divides $Q(x, y)$.

9.4.1   Generation of the $(\mathcal{P}, k, t)$-polynomial

We begin with some definitions.

Definition 9.4.1 For a bivariate polynomial $Q(x, y) = \sum_{i,j} q_{i,j} x^i y^j \in \mathbf{F}_q[x, y]$, its *x-degree*, denoted $\deg_x(Q)$, is defined as the largest integer i with $q_{i,j} \ne 0$, and its *y-degree*, denoted $\deg_y(Q)$, is defined as the largest integer j with $q_{i,j} \ne 0$.

Example 9.4.2 Let $q = 2$ and

$$Q(x, y) = (x + x^4) + (1 + x^4)y + (1 + x)y^2.$$

Then, $\deg_x(Q) = 4$ and $\deg_y(Q) = 2$.

Definition 9.4.3 For an integer $r > 0$, a pair $(\alpha, \beta) \in \mathbf{F}_q^2$ is called an *r-singular point* of $Q(x, y) \in \mathbf{F}_q[x, y]$ if the coefficients of the polynomial $Q(x + \alpha, y + \beta) = \sum_{i,j} q'_{i,j} x^i y^j$ satisfy $q'_{i,j} = 0$, for all i, j with $i + j < r$.

Example 9.4.4 Let $q = 2$ and let $Q(x, y)$ be as in Example 9.4.2. Consider the pair $(1, 1) \in \mathbf{F}_2^2$. It can be checked easily that

$$Q(x + 1, y + 1) = x^4 y + x y^2,$$

so $(1, 1)$ is a 3-singular point of $Q(x, y)$.

Lemma 9.4.5 *Assume that $(\alpha, \beta) \in \mathbf{F}_q^2$ is an r-singular point of $Q(x, y) \in \mathbf{F}_q[x, y]$. Then, for any $f(x) \in \mathbf{F}_q[x]$ with $f(\alpha) = \beta$, $(x - \alpha)^r$ divides $Q(x, f(x))$.*

Proof. Since $(\alpha, \beta) \in \mathbf{F}_q^2$ is an r-singular point of $Q(x, y)$, x^r divides $Q(x + \alpha, xy + \beta)$, and thus $(x - \alpha)^r$ divides $Q(x, (x - \alpha)g(x) + \beta)$, for any $g(x) \in \mathbf{F}_q[x]$. As $f(\alpha) = \beta$, we have that $x - \alpha$ divides $f(x) - \beta$, so $f(x) = (x - \alpha)g(x) + \beta$, for some $g(x) \in \mathbf{F}_q[x]$. Hence, $(x - \alpha)^r$ divides $Q(x, (x - \alpha)g(x) + \beta) = Q(x, f(x))$. $\square$

Definition 9.4.6 A polynomial $f(x) \in \mathbf{F}_q[x]$ is called a *y-root* of $Q(x, y) \in \mathbf{F}_q[x, y]$ if $Q(x, f(x))$ is identically zero, i.e., $y - f(x)$ divides $Q(x, y)$.

Lemma 9.4.7 *If all the pairs in $\mathcal{P}$ are r-singular points of $Q(x, y) \in \mathbf{F}_q[x, y]$, which satisfies*

$$\deg_x(Q) + k \deg_y(Q) < rt, \tag{9.9}$$

then each polynomial in $\Omega(\mathcal{P}, k, t)$ is a y-root of $Q(x, y)$.

Proof. Assume that $f(x)$ belongs to $\Omega(\mathcal{P}, k, t)$. From $f(x) \in \mathbf{F}_q[x]$, with $\deg(f(x)) \leq k$, and $\deg_x(Q) + k \deg_y(Q) < rt$, we see that $Q(x, f(x))$ is a polynomial of degree at most $rt - 1$. From Lemma 9.4.5, $(x - \alpha_i)^r$ divides $Q(x, f(x))$ for at least t distinct indices i. Hence, $Q(x, f(x))$ is identically zero, i.e., $y - f(x)$ divides $Q(x, y)$. $\qquad\qquad\square$

Lemma 9.4.8 *If m, ℓ are nonnegative integers that satisfy $m < k$ and*

$$|\mathcal{P}| \binom{r+1}{2} < \frac{(2m + k\ell + 2)(\ell + 1)}{2}, \tag{9.10}$$

where $|\mathcal{P}|$ is the cardinality of $\mathcal{P}$, then there exists at least one nonzero bivariate polynomial $Q(x, y) \in \mathbf{F}_q[x, y]$, satisfying

$$\deg_x(Q) + k \deg_y(Q) \leq m + k\ell, \tag{9.11}$$

such that all the pairs in $\mathcal{P}$ are r-singular points of $Q(x, y)$.

Proof. By Exercise 9.12, the pairs in $\mathcal{P}$ are r-singular points of a bivariate polynomial $Q(x, y) = \sum_{i,j} q_{i,j} x^i y^j \in \mathbf{F}_q[x, y]$ if and only if the constraint

$$\sum_{i' \geq i, j' \geq j} \binom{i'}{i} \binom{j'}{j} q_{i', j'} \alpha^{i'-i} \beta^{j'-j} = 0 \tag{9.12}$$

holds for all the pairs $(\alpha, \beta) \in \mathcal{P}$ and for all the nonnegative integers i, j with $i + j < r$. The number of constraints of the form (9.12) is equal to $|\mathcal{P}| \binom{r+1}{2}$. From (9.11) and $m < k$, the number of unknowns in the constraints (9.12) is equal to

$$\sum_{j=0}^{\ell} \sum_{i=0}^{(\ell-j)k+m} 1 = \frac{(2m + k\ell + 2)(\ell + 1)}{2}. \tag{9.13}$$

Thus, from (9.10) and from the fact that the constraints (9.12) are linear in the unknowns, we conclude that a nonzero bivariate polynomial $Q(x, y) = \sum_{i,j} q_{i,j} x^i y^j$, satisfying the constraints (9.12), does exist. $\qquad\square$

Definition 9.4.9 A sequence (ℓ, m, r) of nonnegative integers is called a $(\mathcal{P}, k, t)$-*sequence* if $m < \min\{k, rt - \ell k\}$ and (9.10) holds.

Theorem 9.4.10 *If* $t > \sqrt{k|\mathcal{P}|}$, *a* $(\mathcal{P}, k, t)$-*polynomial* $Q(x, y) = \sum_{i,j} q_{i,j} x^i y^j \in$ $\mathbf{F}_q[x, y]$ *with* $\deg_y(Q) = O(\sqrt{k|\mathcal{P}|^3})$ *can be found in polynomial time by solving a linear system whose constraints are of the form* (9.12).

Proof. Let (ℓ, m, r) be the $(\mathcal{P}, k, t)$-sequence given in Exercise 9.13. From Lemma 9.4.8, a nonzero bivariate polynomial $Q(x, y)$ satisfying (9.11) can be found in polynomial time by solving a linear system with constraints of the form (9.12).

Since $m + \ell k < rt$, it follows from (9.11) and Lemma 9.4.7 that all the polynomials in $\Omega(\mathcal{P}, k, t)$ are y-roots of $Q(x, y)$. Hence, $Q(x, y)$ is a $(\mathcal{P}, k, t)$-polynomial.

By the choice of r in Exercise 9.13, we see that $r = O(k|\mathcal{P}|/(t^2 - k|\mathcal{P}|))$. Therefore,

$$\deg_y(Q) \leq \ell = O(t|\mathcal{P}|/(t^2 - k|\mathcal{P}|)). \tag{9.14}$$

Let t_0 be the smallest integer such that $t_0^2 - k|\mathcal{P}| \geq 1$. Then $t_0 = O(\sqrt{k|\mathcal{P}|})$. Now, $t/(t^2 - k|\mathcal{P}|)$ is monotone decreasing in t for $t > \sqrt{k|\mathcal{P}|}$, so it follows from (9.14) that

$$\deg_y(Q) = O(|\mathcal{P}|\sqrt{k|\mathcal{P}|}) = O(\sqrt{k|\mathcal{P}|^3}).$$

This completes the proof of Theorem 9.4.10. □

Remark 9.4.11 The y-degree of a $(\mathcal{P}, k, t)$-polynomial can serve as an upper bound for the cardinality of $\Omega(\mathcal{P}, k, t)$.

9.4.2 Factorization of the $(\mathcal{P}, k, t)$-polynomial

To reconstruct the set $\Omega(\mathcal{P}, k, t)$, it is enough to find all the y-roots $f(x) \in \mathbf{F}_q[x]$, with $\deg(f(x)) \leq k$, of a $(\mathcal{P}, k, t)$-polynomial $Q(x, y)$. Since many efficient algorithms for factorizing univariate polynomials over $\mathbf{F}_q$ are available in the literature (see, for example, Chap. 3 of ref. [14]), we do not discuss here the factorization of such polynomials.

Lemma 9.4.12 *Assume that* $f_0(x) = \sum_{i \geq 0} a_i x^i \in \mathbf{F}_q[x]$ *is a* y-*root of a nonzero bivariate polynomial* $Q_0(x, y)$. *Let* $Q_0^*(x, y) = Q_0(x, y)/x^{\sigma_0}$ *and* $Q_1(x, y) = Q_0^*(x, xy + a_0)$, *where* σ_0 *is the largest integer such that* x^{σ_0} *divides* $Q_0(x, y)$. *Then, a_0 is a root of the nonzero univariate polynomial* $Q_0^*(0, y)$, *and*

$f_1(x) = \sum_{i \geq 0} a_{i+1} x^i \in F_q[x]$ *is a y-root of the nonzero bivariate polynomial* $Q_1(x, y)$.

Proof. From the definition, we see easily that both $Q_0^*(0, y)$ and $Q_1(x, y)$ are nonzero polynomials. Since $f_0(x)$ is a y-root of $Q_0(x, y)$, it means that $Q_0(x, f_0(x))$ is identically zero. Then, we have

$$Q_0^*(x, f_0(x)) = Q_0(x, f_0(x))/x^{\sigma_0} = 0, \tag{9.15}$$

and thus $Q_0^*(0, a_0) = Q_0^*(0, f_0(0)) = 0$; i.e., a_0 is a root of $Q_0^*(0, y)$.

From (9.15), we also have that $Q_1(x, f_1(x)) = Q_0^*(x, f_0(x)) = 0$; i.e., $f_1(x)$ is a y-root of $Q_1(x, y)$. □

Lemma 9.4.13 *Assume that* $Q_0^*(x, y) \in F_q[x, y]$ *is a nonzero bivariate polynomial and that* $\alpha \in F_q$ *is a root of multiplicity* h *of* $Q_0^*(0, y)$. *Let* $Q_1(x, y) = Q_0^*(x, xy + \alpha)$ *and let* $Q_1^*(x, y) = Q_1(x, y)/x^{\sigma_1}$, *where* σ_1 *is the largest integer such that* x^{σ_1} *divides* $Q_1(x, y)$. *Then the degree of the univariate polynomial* $Q_1^*(0, y)$ *is at most* h.

Proof. We assume that

$$G(x, y) := Q_0^*(x, y + \alpha) = \sum_{i \geq 0} g_i(x) y^i. \tag{9.16}$$

Since α is a root of $Q_0^*(0, y)$ of multiplicity h, 0 is a root of multiplicity h of $G(x, y)$. Thus $g_i(0) = 0$, for $i = 0, 1, \ldots, h - 1$, and $g_h(0) \neq 0$. Then, from $Q_1(x, y) = G(x, xy)$, we know that x divides $G(x, xy)$ but x^{h+1} does not. Hence, $1 \leq \sigma_1 \leq h$. It follows from $Q_1^*(x, y) = Q_1(x, y)/x^{\sigma_1} = G(x, xy)/x^{\sigma_1}$ and (9.16) that

$$Q_1^*(x, y) = \sum_{i=0}^{\sigma_1} \frac{g_i(x) x^i}{x^{\sigma_1}} y^i + \sum_{i \geq \sigma_1 + 1} g_i(x) x^{i - \sigma_1} y^i. \tag{9.17}$$

Hence, $Q_1^*(0, y)$ is a univariate polynomial of degree at most $\sigma_1 \leq h$. □

For a nonzero bivariate polynomial $Q_0(x, y) \in F_q[x, y]$ and a positive integer j, let $S_j(Q_0)$ denote the set of sequences $(a_0, a_1, \ldots, a_{j-1}) \in F_q^j$ such that a_i is a root of $Q_i^*(0, y)$, for $i = 0, 1, \ldots, j-1$, where $Q_i^*(x, y) = Q_i(x, y)/x^{\sigma_i}$ with x^{σ_i} exactly dividing $Q_i(x, y)$, and $Q_{i+1}(x, y) = Q_i^*(x, xy + a_i)$. Applying Lemmas 9.4.12 and 9.4.13, we obtain the following theorem.

Theorem 9.4.14 *For any nonzero bivariate polynomial* $Q(x, y) \in F_q[x, y]$ *and any positive integer* j, *the cardinality of* $S_j(Q)$ *is at most* $\deg_y(Q)$, *and, for each y-root* $f(x) = \sum_{i \geq 0} a_i x^i \in F_q[x]$ *of* $Q(x, y)$, *the sequence* $(a_0, a_1, \ldots, a_{j-1})$ *belongs to* $S_j(Q)$.

Example 9.4.15 Let δ be a primitive element in $\mathbf{F}_8$ satisfying $\delta^3 + \delta + 1 = 0$. (Unlike in the earlier parts of this chapter, we do not use α to denote a primitive root here as α has already been used for other purposes in this section.) Find all the y-roots $f(x) \in \mathbf{F}_8[x]$, with $\deg(f(x)) \leq 3$, of the following bivariate polynomial:

$$Q(x, y) = (\delta^5 x + \delta^2 x^3 + \delta^6 x^4 + \delta^2 x^5 + \delta^5 x^6 + \delta^2 x^7 + x^8)$$
$$+ (\delta^4 + \delta^3 x^2 + \delta^5 x^4)y + (\delta x + \delta^4 x^3 + \delta^2 x^4)y^2 + y^3.$$

Solution. We have $Q_0^*(x, y) = Q(x, y)$ and $Q_0^*(0, y) = \delta^4 y + y^3$, which has two roots 0 and δ^2. The multiplicity of the latter root is equal to 2.

Case 1: For the root 0 of $Q_0^*(0, y)$, we have

$$Q_1^*(x, y) = Q_0^*(x, xy)/x$$
$$= (\delta^5 + \delta^2 x^2 + \delta^6 x^3 + \delta^2 x^4 + \delta^5 x^5 + \delta^2 x^6 + x^7)$$
$$+ (\delta^4 + \delta^3 x^2 + \delta^5 x^4)y + (\delta x^2 + \delta^4 x^4 + \delta^2 x^5)y^2 + x^2 y^3$$

and $Q_1^*(0, y) = \delta^5 + \delta^4 y$, which has the unique root δ. Then

$$Q_2^*(x, y) = Q_1^*(x, xy + \delta)/x$$
$$= (\delta x + \delta^6 x^2 + \delta^2 x^3 + x^4 + \delta^2 x^5 + x^6) + (\delta^4 + \delta x^2 + \delta^5 x^4)y$$
$$+ (\delta^4 x^5 + \delta^2 x^6)y^2 + x^4 y^3$$

and $Q_2^*(0, y) = \delta^4 y$, which has the unique root 0. Hence,

$$Q_3^*(x, y) = Q_2^*(x, xy)/x$$
$$= (\delta + \delta^6 x + \delta^2 x^2 + x^3 + \delta^2 x^4 + x^5) + (\delta^4 + \delta x^2 + \delta^5 x^4)y$$
$$+ (\delta^4 x^6 + \delta^2 x^7)y^2 + x^6 y^3$$

and $Q_3^*(0, y) = \delta + \delta^4 y$, whose unique root is δ^4. Thus, we have $(0, \delta, 0, \delta^4) \in S_3(Q)$.

Case 2: For the root δ^2 of $Q_0^*(0, y)$, we have

$$Q_1^*(x, y) = Q_0^*(x, xy + \delta^2)/x^2$$
$$= (\delta^5 + \delta^4 x + x^2 + \delta^2 x^3 + \delta^5 x^4 + \delta^2 x^5 + x^6) + (\delta^3 x + \delta^5 x^3)y$$
$$+ (\delta^2 + \delta x + \delta^4 x^3 + \delta^2 x^4)y^2 + xy^3$$

and $Q_1^*(0, y) = \delta^5 + \delta^2 y^2$, which has a root δ^5 (of multiplicity 2). Then

$$Q_2^*(x, y) = Q_1^*(x, xy + \delta^5)/x^2$$
$$= (1 + \delta^4 x + \delta^2 x^3 + x^4) + \delta^5 x^2 y$$
$$+ (\delta^2 + \delta^6 x + \delta^4 x^3 + \delta^2 x^4)y^2 + x^2 y^3$$

and $Q_2^*(0, y) = 1 + \delta^2 y^2$, which has a root δ^6 (of multiplicity 2). Hence,

$$Q_3^*(x, y) = Q_2^*(x, xy + \delta^6)/x^2$$
$$= (\delta^2 + \delta^6 x + \delta^6 x^2 + \delta^4 x^3 + \delta^2 x^4) y^2 + x^3 y^3$$

and $Q_3^*(0, y) = \delta^2 y^2$, which has a root 0 (of multiplicity 2). Thus, we have $(\delta^2, \delta^5, \delta^6, 0) \in S_3(Q)$.

We have just shown that $S_3(Q) = \{(0, \delta, 0, \delta^4), (\delta^2, \delta^5, \delta^6, 0)\}$. The polynomials related to the sequences in $S_3(Q)$ are

$$f(x) = \delta x + \delta^4 x^3 \quad \text{and} \quad g(x) = \delta^2 + \delta^5 x + \delta^6 x^2.$$

Since $y - g(x)$ divides $Q(x, y)$ but $y - f(x)$ does not, the bivariate polynomial $Q(x, y)$ has a unique y-root $g(x) = \delta^2 + \delta^5 x + \delta^6 x^2$ of degree ≤ 3 in $\mathbf{F}_8[x]$.

Indeed, we can also show that $S_4(Q) = \{(0, \delta, 0, \delta^4, \delta^2), (\delta^2, \delta^5, \delta^6, 0, 0)\}$ and then find that $h(x) = \delta x + \delta^4 x^3 + \delta^2 x^4 \in \mathbf{F}_8[x]$, with $\deg(h(x)) \leq 4$, is also a y-root of $Q(x, y)$. Furthermore, from $Q(x, y)/((y - g(x))(y - h(x))) = y - g(x)$, we see that $Q(x, y)$ can be factorized as $Q(x, y) = (y - g(x))^2(y - h(x))$.

According to Theorem 9.4.14, the following recursive factoring algorithm computes all the y-roots $f(x) \in \mathbf{F}_q[x]$ of degree $\leq k$ of a bivariate polynomial $Q(x, y)$ with the help of any factoring algorithm of a univariate polynomial over $\mathbf{F}_q$.

Factoring algorithm

Input: A nonzero bivariate polynomial $Q(x, y) \in \mathbf{F}_q[x, y]$ and a positive integer k.

Output: The set Ω of y-roots $f(x) \in \mathbf{F}_q[x]$, of degree $\leq k$, of $Q(x, y)$.

Step 1: Define $Q_\eta^*(x, y) := Q(x, y)/x^\sigma$, where η denotes a sequence of length 0 and σ is the number such that x^σ exactly divides $Q(x, y)$. Set $j \leftarrow 2$, S as the set of the roots of $Q_\eta^*(0, y)$, $S' \leftarrow \emptyset$ and goto Step 2.

Step 2: For each $\mathbf{s} = (\mathbf{s}', \alpha) \in S$, do
   (i) define $Q_{\mathbf{s}}^*(x, y) := Q_{\mathbf{s}'}^*(x, xy + \alpha)/x^\sigma$, where σ is the number such that x^σ exactly divides $Q_{\mathbf{s}}^*(x, xy + \alpha)$;
   (ii) factorize $Q_{\mathbf{s}}^*(0, y)$ and, for each root β, add $(\mathbf{s}, \beta)$ into S'.
   Goto Step 3.

Step 3: If $j = k + 1$, goto Step 4.

Else, set $j \leftarrow j + 1$, $S \leftarrow S'$, $S' \leftarrow \emptyset$ and goto Step 2.

Step 4: Output the set Ω of polynomials $f(x) = \sum_{i=0}^{k} a_i x^i \in$ $\mathbf{F}_q[x]$, with degree $\leq k$, for which $(a_0, a_1, \ldots, a_k) \in S'$ and $y - f(x)$ divides $Q(x, y)$.

END.

Remark 9.4.16 The above factoring algorithm can be speeded up to some extent by using the result in Exercise 9.16.

Exercises

9.1 Show that $GRS_k(\alpha, \mathbf{v}) = GRS_k(\alpha, \mathbf{w})$ if and only if $\mathbf{v} = \lambda \mathbf{w}$ for some $\lambda \in \mathbf{F}_q^*$.

9.2 Let

$$G = \begin{pmatrix} v_1 & v_2 & \cdots & v_n \\ v_1\alpha_1 & v_2\alpha_2 & \cdots & v_n\alpha_n \\ \vdots & \vdots & \ddots & \vdots \\ v_1\alpha_1^{k-1} & v_2\alpha_2^{k-1} & \cdots & v_n\alpha_n^{k-1} \end{pmatrix}$$

be a generator matrix for the generalized RS code $GRS_k(\alpha, \mathbf{v})$ and let C be the code with generator matrix $(G|\mathbf{u}^{\mathrm{T}})$, where $\mathbf{u} = (0, \ldots, 0, u)$, for some $u \in \mathbf{F}_q^*$. Let $\mathbf{v}' = (v_1', \ldots, v_n')$ be such that $GRS_{n-k}(\alpha, \mathbf{v}')$ is the dual of $GRS_k(\alpha, \mathbf{v})$.

(i) Show that there is some $w \in \mathbf{F}_q^*$ such that $\sum_{i=1}^{n} v_i v_i' \alpha_i^{n-1} + uw = 0$.

(ii) Show that

$$H' = \begin{pmatrix} v_1' & v_2' & \cdots & v_n' & 0 \\ v_1'\alpha_1 & v_2'\alpha_2 & \cdots & v_n'\alpha_n & 0 \\ v_1'\alpha_1^2 & v_2'\alpha_2^2 & \cdots & v_n'\alpha_n^2 & 0 \\ \vdots & \vdots & \ddots & \vdots \\ v_1'\alpha_1^{n-k} & v_2'\alpha_2^{n-k} & \cdots & v_n'\alpha_n^{n-k} & w \end{pmatrix}$$

is a parity-check matrix for C.

(iii) Show that any $n - k + 1$ columns of H' are linearly independent.

(iv) Prove that C is an MDS code.

9.3 Let

$$A = \begin{pmatrix} a_{11} & a_{12} & \cdots & a_{1t} \\ a_{21} & a_{22} & \cdots & a_{2t} \\ \vdots & \vdots & \ddots & \vdots \\ a_{t1} & a_{t2} & \cdots & a_{tt} \end{pmatrix}$$

be an invertible matrix, where $a_{ij} \in F_{q^m}$, for all $1 \le i, j \le t$. For $1 \le i \le t$, let

$$f_i(x) = a_{i1} + a_{i2}x + a_{i3}x^2 + \cdots + a_{it}x^{t-1}.$$

Show that, for $c \in F_q^n$, $\alpha = (\alpha_1, \ldots, \alpha_n)$ and $v' = (v'_1, \ldots, v'_n)$, we have $c \in A_{n-t}(\alpha, v)$ if and only if $cH'^T = 0$, where

$$H' = \begin{pmatrix} v'_1 f_1(\alpha_1) & v'_2 f_1(\alpha_2) & \cdots & v'_n f_1(\alpha_n) \\ v'_1 f_2(\alpha_1) & v'_2 f_2(\alpha_2) & \cdots & v'_n f_2(\alpha_n) \\ \vdots & & \ddots & \vdots \\ v'_1 f_t(\alpha_1) & v'_2 f_t(\alpha_2) & \cdots & v'_n f_t(\alpha_n) \end{pmatrix}.$$

(Note: this is the reason for the name 'alternant code', as a matrix or determinant of the form

$$\begin{pmatrix} f_1(\alpha_1) & f_2(\alpha_1) & \cdots & f_t(\alpha_1) \\ f_1(\alpha_2) & f_2(\alpha_2) & \cdots & f_t(\alpha_2) \\ \vdots & & \ddots & \vdots \\ f_1(\alpha_n) & f_2(\alpha_n) & \cdots & f_t(\alpha_n) \end{pmatrix}$$

is called an alternant.)

9.4 Let $\gcd(n, q) = 1$ and let F_{q^m} be the smallest extension of F_q containing all the nth roots of 1. Let α be a primitive nth root of 1 in F_{q^m}, so $\{1, \alpha, \ldots, \alpha^{n-1}\} \subseteq F_{q^m}$ are all the nth roots of 1. For $c(x) = \sum_{i=0}^{n-1} c_i x^i \in F_q[x]$, let $\hat{c}(z) \in F_{q^m}[z]$ be defined by

$$\hat{c}(z) = \sum_{j=1}^{n} \hat{c}_j z^{n-j}, \quad \text{where} \quad \hat{c}_j = c(\alpha^j) = \sum_{i=0}^{n-1} c_i \alpha^{ij}.$$

(Note: the polynomial $\hat{c}(z)$ is called the *Mattson–Solomon polynomial* or the *discrete Fourier transform* of $c(x)$.)

(i) Show that $c(x) = \dfrac{1}{n} \sum_{i=0}^{n-1} \hat{c}(\alpha^i) x^i$.

(ii) For a polynomial $f(x)$, recall that $(f(x) \pmod{x^n - 1})$ denotes the remainder when $f(x)$ is divided by $x^n - 1$. For polynomials

$f(x) = \sum_{i=0}^{n-1} f_i x^i$ and $g(x) = \sum_{i=0}^{n-1} g_i x^i$, let

$$f(x) * g(x) = \sum_{i=0}^{n-1} f_i g_i x^i.$$

(a) Show that $(\widehat{f + g})(z) = \hat{f}(z) + \hat{g}(z)$.
(b) Show that $h(x) = (f(x)g(x)) \,(\mathrm{mod}\; x^n - 1))$ if and only if $\hat{h}(z) = \hat{f}(z) * \hat{g}(z)$.
(c) Show that $\hat{h}(z) = \dfrac{1}{n}(\hat{f}(z)\hat{g}(z) \,(\mathrm{mod}\; z^n - 1))$ if and only if $h(x) = f(x) * g(x)$.

9.5 Let the notation be as in Exercise 9.4. Let $\hat{f}(z), \hat{g}(z) \in \mathbf{F}_{q^m}[z]$ be polynomials relatively prime to $z^n - 1$ with $\deg(\hat{f}(z)) \le n - 1$ and $t = \deg(\hat{g}(z)) \le n - 1$. Let $GBCH(\hat{f}, \hat{g})$ be defined as

$$GBCH(\hat{f}, \hat{g}) = \{(c_0, \ldots, c_{n-1}) \in \mathbf{F}_q^n \;:\; (\hat{c}(z)\hat{f}(z) \,(\mathrm{mod}\; z^n - 1))$$
$$\equiv 0 \,(\mathrm{mod}\; \hat{g}(z))\},$$

where $c(x) = \sum_{i=0}^{n-1} c_i x^i$. Let $f(x), g(x) \in \mathbf{F}_q[x]$ be such that $\hat{f}(z), \hat{g}(z)$ are their respective Mattson–Solomon polynomials.

(i) Show that, if $f(x) = \sum_{i=0}^{n-1} f_i x^i$ and $g(x) = \sum_{i=0}^{n-1} g_i x^i$, then $f_i \ne 0$ and $g_i \ne 0$, for all $0 \le i \le n - 1$.
(ii) Show that the following conditions are equivalent:
  (a) $\mathbf{c} = (c_0, \ldots, c_{n-1}) \in GBCH(\hat{f}, \hat{g})$;
  (b) there is a polynomial $\hat{u}(z)$ with $\deg(\hat{u}(z)) \le n - t - 1$ such that

$$(\hat{c}(z)\hat{f}(z) \,(\mathrm{mod}\; z^n - 1)) = \hat{u}(z)\hat{g}(z);$$

  (c) there is a polynomial $u(x) \in \mathbf{F}_{q^m}[x]$ such that $c(x) * f(x) = u(x) * g(x)$ and $\hat{u}_j = 0$ for $1 \le j \le t$, where $\hat{u}(z) = \sum_{j=1}^{n} \hat{u}_j z^{n-j}$ is the Mattson–Solomon polynomial of $u(x) = \sum_{i=0}^{n-1} u_i x^i$;
  (d) there exist $u_0, \ldots, u_{n-1} \in \mathbf{F}_{q^m}$ such that $c_i f_i = u_i g_i$, for $0 \le i \le n - 1$, and $\hat{u}_j = 0$, for $1 \le j \le t$;
  (e) $\hat{u}_j = \sum_{i=0}^{n-1} c_i f_i \alpha^{ij} / g_i = 0$, for all $1 \le j \le t$.
(iii) Show that $\mathbf{c} \in GBCH(\hat{f}, \hat{g})$ if and only if $\mathbf{c}H^{\mathrm{T}} = \mathbf{0}$, where H is equal to

$$\begin{pmatrix} 1 & 1 & 1 & \cdots & 1 \\ 1 & \alpha & \alpha^2 & \cdots & \alpha^{n-1} \\ \vdots & \vdots & & \ddots & \vdots \\ 1 & \alpha^{t-1} & \cdots & \cdots & \alpha^{(t-1)(n-1)} \end{pmatrix} \begin{pmatrix} f_0/g_0 & 0 & \cdots & 0 \\ 0 & f_1\alpha/g_1 & \cdots & 0 \\ \vdots & & \ddots & \vdots \\ 0 & \cdots & 0 & f_{n-1}\alpha^{n-1}/g_{n-1} \end{pmatrix}.$$

(Note: therefore, $GBCH(\hat{f}, \hat{g})$ is an alternant code. It is called a *Chien–Choy generalized BCH code*.)

9.6 Let n be odd and let $\mathbf{F}_{2^m}$ be an extension of $\mathbf{F}_2$ containing all the nth roots of 1. Let α be a primitive nth root of 1 in $\mathbf{F}_{2^m}$ and let $L = \{1, \alpha, \ldots, \alpha^{n-1}\}$. For $\mathbf{c} = (c_0, \ldots, c_{n-1}) \in \mathbf{F}_2^n$, let $R_\mathbf{c}(z) = \sum_{i=0}^{n-1} c_i/(z + \alpha^i)$, as in Definition 9.3.1. Let $c(x) = \sum_{i=0}^{n-1} c_i x^i$ and let $\hat{c}(z)$ be its Mattson–Solomon polynomial.

(i) Show that $\hat{c}(z) = (z(z^n + 1)R_\mathbf{c}(z) \pmod{z^n - 1})$ and

$$R_\mathbf{c}(z) = \sum_{i=0}^{n-1} \frac{\hat{c}(\alpha^i)}{z + \alpha^i}.$$

(ii) Show that the Goppa code $\Gamma(L, g)$ is equal to

$$\Gamma(L, g) = \{\mathbf{c} \in \mathbf{F}_2^n \; : \; (z^{n-1}\hat{c}(z) \pmod{z^n - 1}) \equiv 0 \pmod{g(z)}\}.$$

(Hint: For (i), show that $z(z^n + 1)R_\mathbf{c}(z) = \sum_{i=0}^{n-1} c_i z \prod_{j \neq i}(z + \alpha^j)$. Then show that $(z \prod_{j \neq i}(z + \alpha^j) \pmod{z^n - 1}) = \sum_{j=0}^{n-1} \alpha^{-ij} z^j$ by multiplying both sides by $z + \alpha^i$. For (ii), show that $\mathbf{c} \in \Gamma(L, g)$ if and only if $\sum_{i=0}^{n-1} c_i \prod_{j \neq i}(z + \alpha^j) \equiv 0 \pmod{g(z)}$, and then use (i).)

9.7 Let the notation be as in Exercise 9.6 and suppose that $\Gamma(L, g)$ is a cyclic code. Show that $g(z) = z^t$ for some t and, when $n = 2^m - 1$, that $\Gamma(L, g)$ is a BCH code.

9.8 Let $\alpha_1, \ldots, \alpha_n, w_1, \ldots, w_t$ be distinct elements of $\mathbf{F}_{q^m}$ and let $z_1, \ldots, z_n$ be nonzero elements of $\mathbf{F}_{q^m}$. Let $C = \{\mathbf{c} \in \mathbf{F}_q^n \; : \; \mathbf{c}H^T = \mathbf{0}\}$, where

$$H = \begin{pmatrix} z_1/(\alpha_1 - w_1) & z_2/(\alpha_2 - w_1) & \cdots & z_n/(\alpha_n - w_1) \\ z_1/(\alpha_1 - w_2) & z_2/(\alpha_2 - w_2) & \cdots & z_n/(\alpha_n - w_2) \\ \vdots & \vdots & \ddots & \vdots \\ z_1/(\alpha_1 - w_t) & z_2/(\alpha_2 - w_t) & \cdots & z_n/(\alpha_n - w_t) \end{pmatrix}.$$

Show that C is equivalent to a Goppa code. (Note: this code is called a *Srivastava code*.)

9.9 When $m = 1$ in Exercise 9.8, show that the Srivastava code C is MDS.

9.10 Let C be the binary cyclic code of length 15 with $x^2 + x + 1$ as the generator polynomial. Show that C is a BCH code but not a Goppa code.

9.11 Let $L = \mathbf{F}_8$ and let $g(z) = 1 + z + z^2$. Find the extended binary Goppa code $\overline{\Gamma(L, g)}$ and show that it is cyclic.

9.12 Assume that $Q(x, y) = \sum_{i,j} q_{i,j} x^i y^j \in \mathbf{F}_q[x, y]$ and $(\alpha, \beta) \in \mathbf{F}_q^2$. Prove that the coefficients of $Q(x+\alpha, y+\beta) = \sum_{i,j} q'_{i,j} x^i y^j \in \mathbf{F}_q[x, y]$ satisfy

$$q'_{i,j} = \sum_{i' \geq i, j' \geq j} \binom{i'}{i}\binom{j'}{j} q_{i',j'} \alpha^{i'-i} \beta^{j'-j},$$

for all nonnegative integers i and j.

9.13 For $\mathcal{P} \subseteq \mathbf{F}_q^2$, let $\gamma = k|\mathcal{P}|$. Assume that $t > \sqrt{\gamma}$. Prove that (ℓ, m, r) is a $(\mathcal{P}, k, t)$-sequence, where

$$r = 1 + \left\lfloor \frac{\gamma + \sqrt{\gamma^2 + 4(t^2 - \gamma)}}{2(t^2 - \gamma)} \right\rfloor,$$

$$\ell = \left\lfloor \frac{rt - 1}{k} \right\rfloor,$$

$$m = rt - 1 - \ell k.$$

9.14 For every integer k such that $0 < k < n$, find (or design an algorithm to compute) the smallest positive number $T(n, k)$ such that, for any t with $T(n, k) \leq t < n$ and $\mathcal{P} \subseteq \mathbf{F}_q^2$ with $|\mathcal{P}| = n$, there is at least one $(\mathcal{P}, k, t)$-sequence.

9.15 For integers n, k, t satisfying $0 < k < n$ and $T(n, k) \leq t < n$, find (or design an algorithm to compute) the smallest positive number $\ell = L(n, k, t)$ such that, for any set $\mathcal{P} \subseteq \mathbf{F}_q^2$ with $|\mathcal{P}| = n$, there is at least one $(\mathcal{P}, k, t)$-sequence of the form (ℓ, m, r).

9.16 Let $Q_0(x, y) \in \mathbf{F}_q[x, y]$ be a nonzero bivariate polynomial. Assume that $(a_0, a_1, \ldots, a_j) \in S_{j+1}(Q_0)$, $Q_i^*(x, y) = Q_i(x, y)/x^{\sigma_i}$, where x^{σ_i} exactly divides $Q_i(x, y)$, and $Q_{i+1}(x, y) = Q_i^*(x, xy + a_i)$, for $i = 0, 1, \ldots, j$. Prove that

(i) $\sum_{i=0}^{j} a_i x^i$ is a y-root of $Q_0(x, y)$ if and only if $Q_{j+1}(x, 0) = 0$;

(ii) if $a_j = 0$ and there is a positive number h such that

$$Q_j^*(x, y) = \sum_{i \geq h} g_i(x) y^i \text{ and } g_h(0) \neq 0,$$

then, for any $j' > j$ and sequence $(b_0, b_1, \ldots, b_{j'-1}) \in S_{j'}(Q_0)$ with $b_i = a_i$, for all $0 \leq i \leq j$, the equality $b_l = 0$ must hold for all l such that $j < l < j'$.

References

[1] R. C. Bose and D. K. Ray-Chaudhuri, On a class of error-correcting binary group codes, *Inform. Control* **3**, (1960), 68–79.

[2] P. Delsarte, An algebraic approach to coding theory, *Philips Research Reports Supplements* **10** (1973).

[3] V. D. Goppa, A new class of linear error-correcting codes, *Probl. Peredach. Inform.* **6**(3), (1970), 24–30.

[4] V. D. Goppa, Rational representation of codes and (L, g) codes, *Probl. Peredach. Inform.* **7**(3), (1971), 41–49.

[5] D. Gorenstein and N. Zierler, A class of cyclic linear error-correcting codes in p^m symbols, *J. Soc. Ind. App. Math.* **9**, (1961), 107–214.

[6] V. Guruswami and M. Sudan, Improved decoding of Reed-Solomon and algebraic-geometry codes, *IEEE Trans. Inform. Theory* **45**, (1999), 1757–1767.

[7] A. R. Hammons, Jr., P. V. Kumar, A. R. Calderbank, N. J. A. Sloane and P. Solé, The Z_4-linearity of Kerdock, Preparata, Goethals, and related codes, *IEEE Trans. Inform. Theory* **40**, (1994), 301–319.

[8] A. Hocquenghem, Codes correcteurs d'erreurs, *Chiffres* **2**, (1959), 147–156.

[9] A. N. Kolmogorov and S. V. Fomin, *Introductory Real Analysis*, Translated and Edited by Richard A. Silverman, Dover, New York, (1970).

[10] V. Levenshtein, Application of Hadamard matrices to one problem of coding theory, *Problemy Kibernetiki* **5**, (1961), 123–136.

[11] R. Lidl and H. Niederreiter, *Finite Fields*, Addison-Wesley, Reading, MA (1983); now distributed by Cambridge University Press.

[12] J. H. van Lint, A survey of perfect codes, *Rocky Mountain J. Math.* **5**, (1975), 199–224.

[13] F. J. MacWilliams and N. J. A. Sloane, *The Theory of Error-Correcting Codes*, North-Holland, Amsterdam (1998).

[14] M. Mignotte and D. Ştefănescu, *Polynomials: An Algorithmic Approach*, Springer Series in Discrete Mathematics and Theoretical Computer Science, Springer, Singapore (1999).

[15] A. A. Nechaev, Kerdock code in a cyclic form, *Diskretnaya Mat. (USSR)* **1**, (1989) 123–139. English translation: *Discrete Math. Appl.* **1**, (1991), 365–384.

[16] A. W. Nordstrom and J. P. Robinson, An optimal nonlinear code, *Inform. Control* **11**, (1967), 613–616.

[17] E. Prange, Cyclic error-correcting codes in two symbols, *AFCRC-TN-57*, **103** September (1957).

[18] R. M. Roth and G. Ruckenstein, Efficient decoding of Reed-Solomon codes beyond half the minimum distance, *IEEE Trans. Inform. Theory* **46**, (2000), 246–257.

[19] N. V. Semakov and V. A. Zinov'ev, Complete and quasi-complete balanced codes, *Probl. Peredach. Inform.* **5**(2), (1969), 11–13.

[20] R. C. Singleton, Maximum distance q-nary codes, *IEEE Trans. Inform. Theory* **10**, (1964), 116–118.

[21] M. Sudan, Decoding of Reed-Solomon codes beyond the error-correction bound, *J. Complexity* **13**, (1997), 180–193.

[22] A. Tietäväinen, On the nonexistence of perfect codes over finite fields, *SIAM J. Appl. Math.* **24**, (1973), 88–96.

[23] A. Tietäväinen, A short proof for the nonexistence of unknown perfect codes over $GF(q)$, $q > 2$, *Ann. Acad. Sci. Fenn. Ser. A I Math.* **580**, (1974), 1–6.

[24] Z.-X. Wan, *Quaternary Codes*, World Scientific, Singapore (1997).

[25] V. A. Zinov'ev and V. K. Leont'ev, The nonexistence of perfect codes over Galois fields, *Prob. Control and Info. Theory* **2**, (1973), 123–132.

Bibliography

E. R. Berlekamp, *Algebraic Coding Theory*, McGraw-Hill, New York (1968).
Goppa codes, *IEEE Trans. Inform. Theory* **19**, (1973), 590–592.

E. N. Gilbert, A comparison of signalling alphabets, *Bell Syst. Tech. J.* **31**, (1952), 504–522.

M. J. E. Golay, Notes on digital coding, *Proc. IEEE* **37**, (1949), 657. Anent codes, priorities, patents, etc., *Proc. IEEE* **64**, (1976), 572.

R. W. Hamming, Error detecting and error correcting codes, *Bell Syst. Tech. J.* **29**, (1950), 147–160.

H. J. Helgert, Alternant codes, *Info. and Control* **26**, (1974), 369–380.

R. Hill, An extension theorem for linear codes, *Designs, Codes and Crypto.* **17**, (1999), 151–157.

G. Hughes, A Vandermonde code construction, *IEEE Trans. Inform. Theory* **47**, (2001), 2995–2998.

S. Lin and D. J. Costello, Jr., *Error Control Coding: Fundamentals and Applications*, Prentice-Hall, Inc., New Jersey (1983).

S. Ling and P. Solé, On the algebraic structure of quasi-cyclic codes I: finite fields, *IEEE Trans. Inform. Theory* **47**, (2001), 2751–2760.

I. S. Reed and G. Solomon, Polynomial codes over certain finite fields, *J. Soc. Ind. App. Math.* **8**, (1960), 300–304.

S. A. Vanstone and P. C. van Oorschot, *An Introduction to Error Correcting Codes with Applications*, Kluwer Academic Publishers, Dordrecht (1989).

R. R. Varshamov, Estimate of the number of signals in error correcting codes, *Dokl. Akad. Nauk SSSR* **117**, (1957), 739–741.

C. Xing and S. Ling, A class of linear codes with good parameters, *IEEE Trans. Inform. Theory* **46**, (2000), 2184–2188.

217

Index

alphabet, 5
   channel, 6
   code, 5
alternant code, 192

basis, 42
BCH code, 161
   Chien–Choy generalized, 211
   narrow-sense, 161
   primitive, 161
bound
   Gilbert–Varshamov, 82, 107
   Griesmer, 101
   Hamming, 83
   linear programming, 103, 104
   Plotkin, 95, 96
   Reiger, 150
   Singleton, 92
   sphere-covering, 80
   sphere-packing, 83
burst, 150
burst error, 150

Cauchy–Schwarz inequality, 95
channel
   binary symmetric, 7
   q-ary symmetric, 7
   useless, 14
characteristic, 21
check digits, 59
circulant, 186
co-prime, 23
code
   alternant, 192
   BCH, 159
   block, 5

   burst-error-correcting, 150
   concatenated, 121
   constacyclic, 157
   constant-weight binary, 110
   cyclic, 133
   Delsarte–Goethals, 99
   dual, 45
   equivalent, 56
   exactly u-error-correcting, 13
   exactly u-error-detecting, 12
   expurgated, 69
   expurgated QR, 187
   first order Reed–Muller, 121, 118
   generalized Reed–Solomon, 191
   Golay, 91, 92
   Goppa, 196
   Hadamard matrix, 98
   Hamming, 84, 87
   inner, 122
   irreducible cyclic, 157
   irreducible Goppa, 196
   Kerdock, 99
   linear, 39, 45
   l-burst-error-correcting, 150
   ℓ-quasi-cyclic, 157
   MacDonald, 110
   MDS, 93
   negacyclic, 157
   nonlinear, 96
   Nordstrom–Robinson, 98
   optimal, 76
   outer, 122
   perfect, 84
   Preparata, 99
   punctured, 79
   QR, 180